NanoScience and Technology

Springer
Berlin
Heidelberg
New York
Barcelona
Budapest
Hong Kong
London
Milan
Paris
Santa Clara
Singapore
Tokyo

NanoScience and Technology

Series Editors: K. von Klitzing   R. Wiesendanger

Sliding Friction
Physical Principles and Applications
By B. N. J. Persson

Scanning Probe Microscopy
Analytical Methods
Editor: R. Wiesendanger

Mesoscopic Physics and Electronics
Editors: T. Ando, Y. Arakawa, K. Furuya, S. Komiyama,
and H. Nakashima

Roland Wiesendanger (Ed.)

Scanning Probe Microscopy

Analytical Methods

With 139 Figures

 Springer

Professor Dr. Roland Wiesendanger
Institut für Angewandte Physik
Universität Hamburg
Jungiusstrasse 11
D-20355 Hamburg, Germany

Series Editors:
Professor Dr., Dres. h. c. Klaus von Klitzing
Max-Planck-Institut für Festkörperforschung, Heisenbergstrasse 1
D-70569 Stuttgart, Germany

Professor Dr. Roland Wiesendanger
Institut für Angewandte Physik, Universität Hamburg, Jungiusstrasse 11
D-20355 Hamburg, Germany

ISSN 1434-4904
ISBN 3-540-63815-6 Springer-Verlag Berlin Heidelberg New York

Library of Congress Cataloging-in-Publication Data: Wiesendanger, R. (Roland), 1961– . Scanning probe microscopy: analytical methods/ R. Wiesendanger. p. cm. – (Nanoscience and technology) Includes bibliographical references (p.) and index. ISBN 3-540-63815-6 (hardcover: alk. paper) 1. Scanning probe microscopy. I. Title. II. Series. QH212.S33W534 1998 502'.8'2–dc21 97-48773

© Springer-Verlag Berlin Heidelberg 1998
Printed in Germany

Typesetting: Data conversion by Adam Leinz, Karlsruhe
Cover concept: eStudio Calamar Steinen
Cover production: *design & production*, Heidelberg

SPIN: 10648088   54/3144 - 5 4 3 2 1 0 – Printed on acid-free paper

Dedicated to
Helmut K. V. Lotsch
on the occasion
of his 65th birthday
for his outstanding contributions as an editor
in advanced physical sciences

Preface

Advances in materials research require increasingly powerful analytical techniques with regard to spatial resolution and sensitivity.

The field of Scanning Tunneling Microscopy (STM) and the related Scanning Probe Methods (SPM) has seen a tremendous progress in recent years due to the increasing demand for imaging, measuring and manipulating materials' properties down to the atomic length scale. While atomic or molecular resolution is nowadays routinely achieved on many different kinds of materials there is a strong need for analytical methods at the nanometer or even atomic scale. This volume is devoted to recent progress in scanning probe microscopy with regard to element-specific imaging of multicomponent systems based on the differences in electronic, magnetic, optical or mechanical properties between different chemical species.

After a brief introduction (Chap. 1), T. A. Jung, F. J. Himpsel, R. R. Schlittler, and J. K. Gimzewski will discuss chemical information from scanning probe microscopy and spectroscopy primarily based on measurements of electronic and mechanical material properties. A special chemical contrast mechanism based on the spatially resolved measurement of the thermovoltage across the tunneling barrier will be discussed in Chap. 3 by R. Möller. Chapter 4 by R. Wiesendanger is devoted to magnetic contrast imaging based on concepts of spin-polarized tunneling microscopy and spectroscopy leading to a distinction between different magnetic ions down to the atomic scale. Photon emission from a scanning tunneling microscope as a result of inelastic electron tunneling processes is another powerful spectroscopic method to distinguish between different chemical species and will be discussed in Chap. 5 by R. Berndt. Chapter 6 by M. Völcker will focus on the interaction of laser light with a scanning tunneling microscopic and its applications towards chemical-specific imaging. Finally, the combination of concepts of scanning probe microscopy and optical spectroscopy will be outlined in detail in Chap. 7 on scanning near-field optical microscopy and spectroscopy by U. C. Fischer.

I would like to thank all contributors for their Chapters to this volume on analytical methods in scanning probe microscopy. I also thank Springer-Verlag for a pleasant and fruitful collaboration. It is hoped that the present volume will put the role of scanning probe microscopy in analytical chemistry into perspective and that it will stimulate further efforts towards a non-destructive atomic-resolution material analysis.

Hamburg, January 1998   *R. Wiesendanger*

Contents

Contributors

R. Berndt
2. Institute of Physics
RWTH Aachen
Templergraben 55
D-52056 Aachen, Germany

U. Fischer
Institute of Physics
University of Münster
Wilhelm-Klemm-Straße 10
D-48149 Münster, Germany

J. K. Gimzewski
IBM Research Division
Zürich Research Laboratory
Säumerstraße 4
CH-8803 Rüschlikon, Switzerland

F. J. Himpsel
Department of Physics
University of Wisconsin at Madison
1150 University Ave.
Madison, WI 53706-1390, USA

T. A. Jung
Laboratory for Micro-
and Nanostructures
Paul Scherrer Institute
CH-5232 Villigen, Switzerland

R. Möller
Universität Gesamthochschule Essen
Fachbereich 7 – Physik
Insitut für Laser-Plasmaphysik
D-45117 Essen, Germany

R. R. Schlittler
IBM Research Division
Zürich Research Laboratory
Säumerstraße 4
CH-8803 Rüschlikon, Switzerland

M. Völcker
Carl Zeiss, FO-IB
D-73446 Oberkochen, Germany

R. Wiesendanger
Institute of Applied Physics
University of Hamburg
Jungiusstraße 11
D-20355 Hamburg, Germany

1. Introduction

R. Wiesendanger

With 3 Figures

Since its invention, Scanning Tunneling Microscopy (STM) [1.1–3] and related Scanning Probe Methods (SPM) have proven useful in many different scientific disciplines including surface science, materials science, surface chemistry, electrochemistry, biology, metrology, etc. [1.4–6], and constitute indispensable tools for the new age of nanotechnology. This is primarily based on the ability to image and locally probe surfaces of metals, semiconductors, and insulators down to the atomic level. While atomic or molecular resolution is nowadays routinely achieved on many different kinds of materials, the analytical capabilities of SPM are not yet fully developed. In favorable cases a distinction between different chemical species has been achieved by STM, Atomic Force Microscopy (AFM) or Scanning Near-field Optical Microscopy (SNOM), but SPM is still far from being a standard analytical tool with the ability to identify elemental species. The determination of chemical species by STM or tunneling spectroscopy is inherently prevented because the valence electronic states accessible by STM cannot unequivocally be related with a particular element in a multicomponent solid. However, considerable progress has been made in recent years to exploit various contrast mechanisms to differentiate between different atomic or molecular species.

A first example was given by atom-selective imaging of compound semiconductors, such as GaAs (Fig. 1.1), based on bias-polarity dependent STM studies [1.7]. Charge transfer from Ga to As results in an occupied electronic state centered at the As sites and an empty state centered at the Ga sites. Consequently, STM selectively images the As atoms at negative sample bias and the Ga atoms at positive sample bias. A similar atom-selective imaging mechanism is found for other III–V semiconductor compounds. For the general case of an arbitrary multicomponent solid, however, the details of charge transfer between different atomic species are usually not known and therefore an assignment of "topographic" features in STM images to particular atomic species cannot be made.

Another example is given by the Si(111)–($\sqrt{3} \times \sqrt{3}$) Al system [1.8–10]. STM images obtained at positive sample bias reveal the ($\sqrt{3} \times \sqrt{3}$) lattice together with defect sites which appear less bright. At negative sample bias, the contrast reverses. Based on the observed dependence of the defect density on Al coverage, it was concluded that the bright protrusions seen at positive bias must correspond to Al adatoms, whereas the defect sites were attributed to Si adatoms substituting for Al in the ($\sqrt{3} \times \sqrt{3}$) structure be-

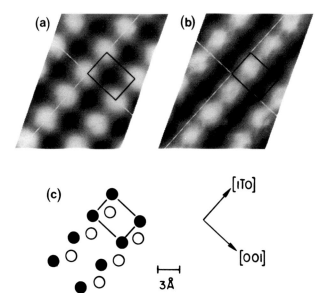

Fig. 1.1. Constant-current STM images of the GaAs(110) surface acquired at sample bias voltages of (a) +1.9 V and (b) −1.9 V. (c) Top-view of the surface atoms. The As atoms are represented by *open circles* and the Ga atoms by *closed circles*. The *rectangle* indicates a unit cell, whose position is the same in all three figures [1.7]

low an Al coverage of 1/3 of a monolayer. The Si adatoms give rise to an extra dangling-bond defect state near −0.4 eV, which causes the Si adatoms to appear brighter than the Al adatoms for negative sample bias. Based on the polarity-dependent contrast, an atom-selective imaging can be achieved for the Si(111)–($\sqrt{3} \times \sqrt{3}$) Al system, as already found for the GaAs (110) surface. However, the assignment of the imaged surface protrusions to specific atomic species again depends on external knowledge and cannot directly be inferred from the STM data.

Alternatively, STM with simultaneous laser excitation has been proposed to achieve chemical contrast on multicomponent semiconductor surfaces down to the atomic level [1.11]. In particular, the use of coordinated electronic and photon biasing has been demonstrated to provide atom-specific imaging for CuInSe$_2$, a covalently bonded semiconductor with applications for thin-film photovoltaic technologies. It is assumed that the photon bias at a particular wavelength directly excites the atom, with the wavelength corresponding to the atomic transition probability. This method has been applied to study the local chemical order in the vicinity of defects, such as grain boundaries. However, the same method has been found to be inappropriate for atom-selective imaging of metal surfaces.

For metal alloys, a distinction between different chemical species appears to be feasible under special tunneling conditions related with the microscopic structure of the probe tip [1.12–16]. For instance, an adsorbed atom at the apex of the tip may tend to form a chemical bond more likely with one element than with the other, which can cause a difference in corrugation amplitudes in constant-current STM traces over the two different chemical species. However, additional information about the tip structure is required in order to attribute the "higher" sites to a particular component of the metal alloy.

Alternatively, STM contrast between different chemical species can be achieved by spatially resolved measurements of the local tunneling barrier height [1.17–19]. This method appears to be useful to study thin film growth or phase separation in multicomponent systems.

More recently, the spectroscopical capabilities of STM have been exploited to differentiate between different chemical species in multicomponent metal systems including binary alloys [1.20], segregated systems [1.21], adsorbate-[1.22] or thin film systems [1.23, 24]. In this case, the chemical contrast is based on characteristic features in the local tunneling spectra arising from surface states, surface resonances or adsorbate-induced states. An example is presented in Fig. 1.2 showing the topography of a submonolayer Fe film ($\theta \approx 0.45$ ML) grown at room temperature on a stepped W(110) substrate. The topography is dominated by iron islands of monolayer height and a lateral extent of about 5 nm, though a small amount of the deposited material tends to decorate the monoatomic steps of the W substrate. A distinction between Fe and W is difficult to make based on such constant-current topographs, particularly close to step edges where the difference in the apparent topographic height between Fe and W almost vanishes. Therefore, chemical-specific fingerprints are needed which can be extracted from the local tunneling spectra measured either above the bare W(110) substrate (Fig. 1.2b, spectrum A) or above the Fe islands (Fig. 1.2b, spectrum B). While the dI/dU–U-characteristic measured above the Fe islands exhibits a pronounced peak centered at $U = +0.2$ V corresponding to an empty d-state, the spectrum

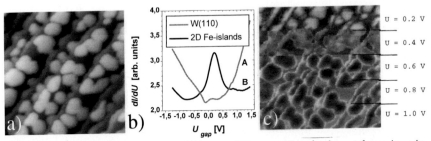

Fig. 1.2. (a) Constant-current STM image (60 nm × 60 nm) of monolayer iron islands on terraces of a stepped W(110) substrate; (b) Local tunneling spectra measured above the W(110) substrate (A) and above the iron islands (B). A pronounced empty-state peak at $U = +0.2$ V appears in spectrum (B). (c) Spatially resolved measurements of the differential tunneling conductivity $dI/dU(x, y)$ for five different values of the applied bias voltage which has been increased by increments of 0.2 V from top to bottom. A contrast reversal is observed in dI/dU (x, y)-images obtained between $U = +0.2$ V and $U = +1.0$ V which can be explained by the pronounced difference of the local tunneling spectra for the iron islands and the W(110) substrate as shown in (b) [1.23, 24]

measured above the bare W(110) substrate shows no significant feature in this bias voltage regime. Based on the pronounced difference between the tunneling spectra measured above Fe and W, a chemical-specific imaging of the ultra-thin film system can be achieved by spatially resolved measurements of the differential tunneling conductivity $dI/dU(x,y)$ for selected values of the applied dc bias voltage. This is demonstrated in Fig. 1.2c, which shows a map of $dI/dU(x,y)$ where the applied bias voltage was changed every 80 scan lines in steps of $\Delta U = 0.2\,V$. At $U = +0.2\,V$ (Fig. 1.2c, top part) the Fe islands appear bright compared to the W substrate corresponding to the high differential tunneling conductivity of Fe at $U = +0.2\,V$ as known from the local tunneling spectrum (Fig. 1.2b, curve B). On the other hand, at $U = +1.0\,V$ (Fig. 1.2c, bottom part) the Fe islands appear dark compared to the W substrate which is explained by the much lower differential tunneling conductivity of Fe compared with W for high sample bias voltage (Fig. 1.2b). Accordingly, a contrast flip in the $dI/dU(x,y)$ image (Fig. 1.2c) is observed between $U = 0.2\,V$ and $U = +1.0\,V$, corresponding to the crossing point of the two local tunneling spectra presented in Fig. 1.2b. To obtain high chemical contrast in the $dI/dU(x,y)$ map, it is therefore important to select an appropriate dc bias voltage for which the difference in the differential tunneling conductivity between different chemical species appears to be the highest. This information can easily be extracted from local dI/dU–U spectra as demonstrated for the Fe/W(110) system in the present case.

The method for chemical-specific imaging of multicomponent metal surfaces at the nanometer scale outlined above is applicable quite generally. It is known that most metal surfaces exhibit some pronounced density-of-states feature within the bias voltage regime accessible by STM. Other examples include Cu(110) for which a pronounced empty electronic state has been identified by inverse photoelectron spectroscopy [1.25], or Cr(001) and Fe(001) for which pronounced density-of-states features close to the Fermi level have been observed by local tunneling spectroscopy [1.26, 20].

Chemical contrast on metal surfaces can also be achieved via spectroscopic STM studies of image states [1.27] as discussed in detail in Chap. 2. In this case, the contrast relies on the work function difference between different elemental species. Another contrast mechanism which has been exploited for "chemical mapping" is based on the thermovoltage across the tunneling barrier which is related with the logarithmic derivative of the electronic density of states. This will be discussed in detail in Chap. 3.

For para-, ferro-, ferri- or antiferromagnetic materials, it is important to be able to distinguish between different magnetic ions on the atomic scale. This has first been demonstrated at the example of Fe_3O_4 (magnetite) for which STM operated with a ferromagnetic Fe probe tip could distinguish between Fe^{2+}- and Fe^{3+}-sites in the (001) plane of octahedrally coordinated Fe sites of Fe_3O_4 [1.28, 29]. The magnetic contrast achieved can be ascribed to spin-polarized tunneling which will be discussed in detail in Chap. 4.

For molecular species inelastic electron tunneling spectroscopy [1.30, 31] combined with STM might become a powerful technique to identify functional groups. A first experiment toward this goal was performed with sorbic acid adsorbed on graphite [1.32]. A series of strong peaks was observed in the tunneling spectrum with the energetic positions of the peaks corresponding approximately to the vibrational modes of the molecule. However, such results are not yet routinely obtained with the STM in contrast to the use of planar tunnel junctions [1.30, 31]. On the other hand, signatures of inelastic electron tunneling events can clearly be observed by the detection of photons emanating from an STM-type tunnel junction [1.33]. This will be discussed in detail in Chap. 5. The combination of STM with optical methods offers many attractive features with regard to spectroscopic contrast mechanisms which may add 'true color' to the STM images. On the other hand, the study of the interaction of laser light with the tunnel junction has led to the development of the laser-driven STM (Chap. 6) and the AC-STM which offer additional opportunities for chemical-selective imaging of materials including insulators. The Scanning Near-field Optical Microscope (SNOM) combined with optical spectroscopy, offers probably the most rich variety of experimental methods for chemical-selective imaging by SPM though the spatial resolution might be limited to the nanometer-scale regime. Chapter 7 will be devoted to this topic.

Finally, the SFM offers various modes of operation which can be exploited for achieving chemical contrast. This includes spatial mapping of the local surface elasticity based on a modulation of the sample height which causes the SFM cantilever to deflect periodically. For a given applied loading force, a soft surface region deforms more than a hard one. Consequently, the measured cantilever deflection is smaller over a soft surface region. Spatial mapping of the local surface elasticity is particularly useful for chemical-selective imaging of composite materials [1.34]. Alternatively, spatial adhesion maps [1.35, 36] extracted from spatially resolved force-versus-distance curves [1.4, 6] can help to distinguish between different chemical species which exhibit different adhesion properties. This method has even been applied to molecular species [1.37–39] for which specific molecular interaction mechanisms based on the existence of geometrically complementary surfaces of recognition sites provide a means to obtain molecular fingerprints. Examples of present SFM studies include molecular recognition between receptor and ligand, antibody and antigen, and complementary strands of DNA [1.37–39].

The detection of lateral or 'frictional' forces in addition to the normal forces acting between tip and sample can also be employed for chemical-selective imaging as demonstrated, for instance, for phase-separated Langmuir-Blodgett films [1.40] However, the contrast mechanism in Frictional Force Microscopy (FFM) critically depends on the applied loading force as first demonstrated at the example of a C_{60} thin film sample [1.41]. A topographic force micrograph of a 1.2 ML thin film of C_{60} on GeS(001) is presented in Fig. 1.3a. The friction force F_f was measured for a series of loads F_l applied

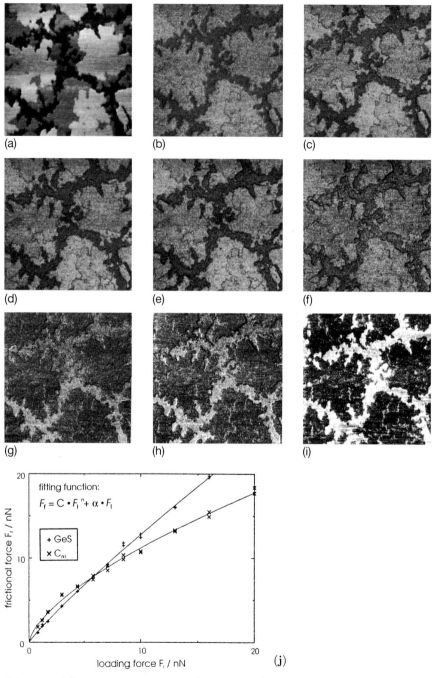

Fig. 1.3. (a) Constant-force image ($2\,\mu$m $\times\,2\,\mu$m) of a $1.2\,$ML thin film of C_{60} on GeS(001); (b)–(i) Corresponding friction force maps for the same area at different loading forces of $1.3\,$nN (b), $2.6\,$nN (c), $4.5\,$nN (d), $6.7\,$nN (e), $10\,$nN (f), $15\,$nN (g), $20\,$nN (h), and $30\,$nN (i). A friction contrast reversal is observed between images (f) and (g) which can be understood by a different friction-load dependence (j) for the C_{60} thin film and the GeS(001) substrate [1.41]

to the area of contact between tip and sample. Friction force maps, which were recorded simultaneously with the topographic image shown in Fig. 1.3a, were obtained for loading forces F_l ranging from 1.3 up to 30 nN (Fig. 1.3b–i). Dark colors in FFM images represent low friction force areas whereas bright colors represent high friction force areas. The FFM maps in Fig. 1.3b–f, obtained with loads below 10 nN, exhibit lower frictional forces on the GeS substrate compared with the C_{60} islands. At approximately 10 nN (Fig. 1.3f) this contrast vanishes and finally flips as the load rises (Fig. 1.3g–i). At high loads, the C_{60} islands therefore exhibit lower frictional forces compared with the GeS substrate. It is clear that the 'chemical contrast' achieved in this model case critically depends on an external parameter, the applied loading force, and that an assignment of chemical species is not possible without understanding the details of the tip-sample contact mechanics. To study the origin of the observed friction contrast reversal as a function of the applied load in more detail, the quantitative dependence of the frictional force on the applied loading force for the GeS substrate and the C_{60} islands have to be measured as shown in Fig. 1.3j. The behaviour of the frictional force on the C_{60} islands (2/3-power law dependence) is markedly different compared with the behaviour on the substrate (linear dependence). As a result, the two data sets exhibit a crossing point, which corresponds to the contrast reversal in the spatially resolved measurements.

Consequently, the interpretation of friction force maps in terms of chemical contrast requires a careful analysis of the local friction force spectroscopy data. The situation is analogous to STM or scanning tunneling spectroscopy: the applied loading force as an important external parameter in FFM studies plays a similar role as the applied sample bias voltage in STM. Again, a combination of local spectroscopic measurements with spatially resolved imaging is required in order to obtain data with clear chemical contrast and clear assignment of the chemical species.

Another SFM-derived technique which offers great potential for three-dimensional (3D) chemical analysis is Magnetic Resonance Force Microscopy (MRFM) [1.42–44]. In these experiments, a force signal is generated by modulating the sample magnetization with standard magnetic resonance techniques. Samples of only a few nanograms generate force signals on the order of 10^{-14}–10^{-16} N which can be detected by high-sensitive microfabricated force sensors. Possible applications of MRFM include 3D-imaging of biological specimens. However, it remains to be demonstrated whether atomic or molecular resolution can actually be achieved by using this technique.

References

[1.1]   G. Binnig, H. Rohrer: Helv. Phys. Acta **55**, 726 (1982)
[1.2]   G. Binnig, H. Rohrer, Ch. Gerber, E. Weibel: Phys. Rev. Lett. **49**, 57 (1982)
[1.3]   G. Binnig, H. Rohrer: Rev. Mod. Phys. **59**, 615 (1987)

[1.4]   R. Wiesendanger, H.-J. Güntherodt (eds.): *Scanning Tunneling Microscopy I–III*, 2nd ed., Springer Ser. Surf. Sci. (Springer Berlin, Heidelberg, New York, 1994, 1995, 1996)

[1.5]   J.A. Stroscio, W.J. Kaiser (eds.): *Scanning Tunneling Microscopy* (Academic Press, Boston 1993

[1.6]   R. Wiesendanger: *Scanning Probe Microscopy and Spectroscopy: Methods and Applications* (Cambridge Univ. Press, Cambridge 1994)

[1.7]   R.M. Feenstra, J.A. Stroscio, J. Tersoff, A.P. Fein: Phys. Rev. Lett. **58** 1192 (1987)

[1.8]   R.J. Hamers, J.E. Demuth: Phys. Rev. Lett. **60**, 2527 (1988)

[1.9]   R.J. Hamers: J. Vac. Sci. Technol. B **6**, 1462 (1988)

[1.10] R.J. Hamers: Phys. Rev. B **40**, 1657 (1989)

[1.11] L.L. Kazmerski: J. Vac. Sci. Technol. B **9**, 1549 (1991)

[1.12] D.D. Chambliss, S. Chiang: Surf. Sci. Lett. **264**, L 187 (1992)

[1.13] M. Schmid, H. Stadler, P. Varga: Phys. Rev. Lett. **70**, 1441 (1993)

[1.14] L. Ruan, F. Besenbacher, I. Stensgaard, E. Laegsgaard: Phys. Rev. Lett. **70**, 4079 (1993)

[1.15] F. Besenbacher, E. Laegsgaard, L. Pleth Nielsen, L. Ruan, I. Stensgaard: J. Vac. Sci. Technol. B **12**, 1758 (1994)

[1.16] P.T. Wouda, B.E. Nieuwenhuys, M. Schmid, P. Varga, Surf. Sci. **359**, 17 (1996)

[1.17] G. Binnig, H. Rohrer: Surf. Sci. **126**, 236 (1983)

[1.18] R. Wiesendanger, L. Eng, H.R. Hidber, P. Oelhafen, L. Rosenthaler, U. Staufer, H.-J. Güntherodt: Surf. Sci. **189/190**, 24 (1987)

[1.19] R. Wiesendanger, M. Bode, R. Pascal, W. Allers, U.D. Schwarz: J. Vac. Sci. Technol. A **14**, 1161 (1996)

[1.20] A. Davies, J.A. Stroscio, D.T. Pierce, R.J. Celotta: Phys. Rev. Lett. **76**, 4175 (1996)

[1.21] A. Biedermann, O. Genser, W. Hebenstreit, M. Schmid, J. Redinger, R. Podloucky, P. Varga: Phys. Rev. Lett. **76**, 4179 (1996)

[1.22] M. Bode, R. Pascal, R. Wiesendanger: Z. Phys. B **99**, 143 (1996)

[1.23] M. Bode, R. Pascal, M. Dreyer, R. Wiesendanger: Phys. Rev. B **54**, R8385 (1996)

[1.24] M. Bode, R. Pascal, R. Wiesendanger: Appl. Phys. A **62**, 571 (1996)

[1.25] Y.W. Mo, F.J. Himpsel: Phys. Rev. B **50**, 7868 (1994)

[1.26] J.A. Stroscio, D.T. Pierce, A. Davies, R.J. Celotta: Phys. Rev. Lett. **75**, 2960 (1995)

[1.27] T. Jung, Y.W. Mo, F.J. Himpsel: Phys. Rev. Lett. **74**, 1641 (1995)

[1.28] R. Wiesendanger, I.V. Shvets, D. Bürgler, G. Tarrach, H.-J. Güntherodt, J.M.D. Coey: Europhys. Lett. **19**, 141 (1992)

[1.29] R. Wiesendanger, I.V. Shvets, D. Bürgler, G. Tarrach, H.-J. Güntherodt, J.M.D. Coey, S. Gräser: Science **255**, 583 (1992)

[1.30] T. Wolfram (ed.): *Inelastic Electron Tunneling Spectroscopy* (Springer, Berlin, Heidelberg 1979)

[1.31] P.K. Hansma (ed.): *Tunneling Spectroscopy* (Plenum, New York 1982)

[1.32] D.P.E. Smith, M.D. Kirk, C.F. Quate: J. Chem. Phys. **86**, 6034 (1987)

[1.33] R. Berndt, R. Gaisch, J.K. Gimzewski, B. Reihl, R.R. Schlittler, W.D. Schneider, M. Tschudy: Science **262**, 1425 (1993)

[1.34] P. Maivald, H.J. Butt, S.A.C. Gould, C.B. Prater, B. Drake, J.A. Gurley, V.B. Elings, P.K. Hansma: Nanotechnology **2**, 103 (1991)

[1.35] H.A. Mizes, K.-G. Loh, R.J.D. Miller, S.K. Ahuja, E.F. Grabowski: Appl. Phys. Lett. **59**, 2901 (1991)

[1.36] K.O. van der Werf, C.A.J. Putman, B.G. de Grooth, J. Greve: Appl Phys. Lett. **65**, 1195 (1994)

[1.37] E.-L. Florin, V.T. Moy, H.E. Gaub: Science **264**, 415 (1994)
[1.38] C.D. Frisbie, L.F. Rozsnyai, A. Noy, M.S. Wrighton, C.M. Lieber: Science **265**, 2071 (1994)
[1.39] G.U. Lee, L.A. Chrisey, R.J. Colton: Science **266**, 771 (1994)
[1.40] R.M. Overney, E. Meyer, J. Frommer, D. Brodbeck, R. Lüthi, L. Howald, H.-J. Güntherodt, M. Fujihira, H. Takano, Y. Gotoh: Nature **359**, 133 (1992)
[1.41] U.D. Schwarz, W. Allers, G. Gensterblum, R. Wiesendanger: Phys. Rev. B **52**, 14976 (1995)
[1.42] D. Rugar, C.S. Yannoni, J.A. Sidles: Nature **360**, 563 (1992)
[1.43] D. Rugar, O. Züger, S. Hoen, C.S. Yannoni, H.-M. Vieth, R.D. Kendrick: Science **264**, 1560 (1994)
[1.44] J.A. Sidles, J.L. Garbini, K.J. Bruland, D. Rugar, O. Züger, S. Hoen, C.S. Yannoni: Rev. Mod. Phys. **67**, 249 (1995)

2. Chemical Information from Scanning Probe Microscopy and Spectroscopy

T. A. Jung, F. J. Himpsel, R. R. Schlittler, and J. K. Gimzewski

With 25 Figures

Chemical information has been essential to our understanding of all processes and mechanisms in science, engineering, and technology. The increasing importance of laterally confined nanometer-sized structures and objects on surfaces and at interfaces demands both high spatial resolution and nondestructive analytical techniques. This chapter gives an account of the various approaches that have evolved from Scanning Probe Microscopies (SPM), which provide chemical information. In particular, we will discuss what types of chemical information can be acquired using various SPM imaging and experimentation techniques. Specifically we address elemental and molecular composition, reactivity, bonding, adsorption and desorption, and site specificity. It is notable that the combination of different techniques provides reliable information with spatial resolution on the nanometer scale. Finally, we exemplify these cooperative approaches to surface analytical problems, and outline the potential for future developments.

2.1 Background

Chemical analysis of matter consists of the determination of either elemental or molecular composition. Traditionally, a small amount of material is separated into different phases by physical methods, and the elemental composition of each phase is determined by quantitative chemical reactions. The amount of material required for chemical analysis has decreased continuously as increasingly sensitive analytical tools have become available. Prominent techniques are Nuclear Magnetic Resonance (NMR), Electron Spin Resonance (ESR), Rutherford Back-Scattering (RBS) of ions, X-ray and optical/UV fluorescence, Atomic Absorption Spectroscopy (AAS), Mass Spectrometry (MS), various versions of chromatography, and optical spectroscopy. Ultimate sensitivity is obtained for single-event counting in gas chromatography/MS. However, the detailed analysis of complex mixtures typically involves larger amounts of material and a combination of methods and experimental approaches. At surfaces and interfaces, the relevant chemical composition varies strongly with monolayer depth and lateral dimensions at grain boundaries. In technological microstructures commercial microelectronics operates on 300-nm design scales. Hence, use of a nondestructive technique and a high, locally confined detection sensitivity are fundamental prerequisites. Typical techniques for the analysis of surfaces and interfaces are Secondary Ion Mass Spectroscopy (SIMS), Auger Electron Spectroscopy (AES), X-ray Photoelectron Spectroscopy (XPS), and X-ray Absorption Spectroscopy (XAS). Viewed edge-on, interfaces can also be analyzed by Scanning Transmission Electron Microscopy (STEM) [*Pennycook* and *Jesson* (1990), *Jesson* et al. (1991)], where Electron Energy Loss Spectroscopy (EELS) or fluorescence spectroscopy [EDX, *Reimer* (1985), *Reed* (1975), *Goldstein* et al. (1981)] of core-level excitations provides element analysis [*Pennycook* and *Boatner* (1988)]. CathodoLuminescence (CL) stimulated by the nanometer-sized electron beam of a Scanning Electron Microscope (SEM) is widely used to characterize optical and electronic properties of semiconductors. Figure 2.1 shows a comparison of the relative resolution and sensitivity of a variety of the most commonly used methods. Generally, a trade-off between sensitivity and spatial resolution is ultimately dictated by the number of atoms available for analysis in a dilute sample. Typical lateral resolution limits of conventional methods are of the order of 0.1–1 μm [*Newbury* (1990)]. STEM/EELS, however, is able to probe a region of 0.2–0.3 nm across and 1–3 nm deep [*Browning* et al. (1993), *Muller* et al. (1993), *Batson* (1993, 1996)]. For an introduction to these techniques of surface characterization and analysis, see *Morrison* (1990).

The invention of Scanning Tunneling Microscopy (STM) and related Scanning Probe Microscopies (SPM) with their imaging capabilities down to the

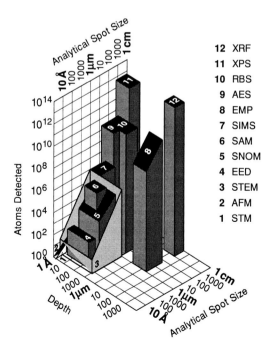

Fig. 2.1. Analytical spot size, information depth and sensitivity of various techniques for chemical analysis. (Adapted from Allen J. Bard, *Integrated Chemical Systems: A Chemical Approach to Nanotechnology* (Wiley-Interscience, New York, 1994) and *Texas Instruments Materials Characterization Capabilities*, Texas Instruments, Dallas, TX, 1986. Courtesy of Texas Instruments; adapted by permission of John Wiley and Sons, Inc.)

atomic range [*Rohrer* (1994), *Quate* (1994)] has raised hopes that it might one day be possible to determine atomic or molecular compositions with a comparable resolution. In the following review we will first show how voltage-dependent tunneling and Scanning Tunneling Spectroscopy (STS) provide such information (Sect. 2.2) via electronic states. Then a variety of other methods to extract chemical information from SPM (Sect. 2.3), will be addressed: Topographic contrast at high resolution, light emission and adsorption confined under a local probe as well as the direct measurement of lateral and perpendicular forces and deformations in SPM provide contrast that depends on chemical factors.

2.2 Contrast via Electronic States

STM senses contours of equal charge density outside the surface produced by electronic states near the Fermi level [*Tersoff* and *Hamann* (1985)]. Although STM images contain both topographical and electronic information, the pictures have commonly been considered representative of atom positions, but not of the chemical nature of the atoms. Other high-resolution microscopies, such as TEM (Transmission Electron Microscopy) and FIM (Field Ion Microscopy) [*Ehrlich* (1992), *Tsong* (1990), *Müller* and *Tsong* (1969)], are in a similar situation: The arrangement of atoms and the geometry of the surface binding sites can be determined, but not their identity. For achieving chemical sensitivity there are several levels of refinement to be considered: First, a contrast mechanism between different atoms is required. Then, their identity must be found, and eventually their chemical bonding can be explored, such as bonding/antibonding orbitals, particularly the HOMO/LUMO orbitals, lone pairs, and surface states in general.

2.2.1 Resonant Tunneling

Chemical sensitivity can be achieved with STM by resonant tunneling into electronic states that are characteristic of the sample surface (Fig. 2.2). Variation of the bias voltage V between sample and tip allows it to place the Fermi level of the tip at a specific energy level of the surface and to emphasize tunneling via that specific state. Positive sample bias corresponds to electrons tunneling from the tip into empty sample states, negative bias to electrons tunneling from occupied sample states into the tip. In both cases the tunnel barrier is lowest for electrons originating from the Fermi level of the emitting electrode (indicated by long, thick arrows in Fig. 2.2). For typical tunnel barriers ΔE of 3–4 eV (comparable to the work functions) the main contribution to the tunneling current comes from electrons emitted within about 0.3 eV of the Fermi level [*Hamers* (1992)]. This can be seen from the energy-dependent transmission probability

$$T(E, V) = \exp\left[\frac{-2z(\Delta E + eV)}{(2 - E)^{1/2}} \frac{(2m)^{1/2}}{\hbar}\right] , \qquad (2.1)$$

Sample bias   **Sample**   **Tip**

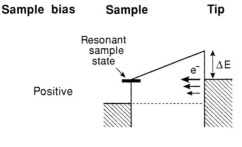

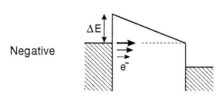

Fig. 2.2. Schematic representation of energy levels for resonant tunneling. Enhanced tunneling from the tip into empty sample states is observed at positive sample bias, whereas sample states at the Fermi level dominate tunneling into the tip at negative sample bias. The energy barrier ΔE is comparable to the work function

where E is the electron energy relative to the Fermi level of the emitter, V the sample bias, z the tip–sample separation, $e > 0$ the elementary charge, and $(2m)^{1/2}/\hbar = 5.1\,\mathrm{nm}^{-1}/(\mathrm{eV})^{1/2}$.

As a consequence, it is possible to tunnel selectively into empty states of the sample at positive sample bias, but not from specific occupied states at positive bias. In the latter case, most of the tunneling current will always originate from the highest occupied state of the sample. Empty states at the surface can be identified independently by inverse photoemission [*Dose* (1985), *Smith* (1988), *Himpsel* (1990b)], a technique that provides a complete set of quantum numbers (energy E, momentum or wave vector $\mathbf{k}$, angular or point group symmetry, and spin). Thereby the elemental parentage of resonant tunneling states can be established.

Bias-dependent tunneling effects are enhanced in the differential measurement mode, where the derivative $\mathrm{d}I/\mathrm{d}V$ of the tunnel current I versus bias voltage V is determined (in most cases at constant tip–sample distance). To a first approximation the quantity

$$\frac{\mathrm{d}I/\mathrm{d}V}{I/V} = \frac{\mathrm{d}(\ln I)}{\mathrm{d}(\ln V)} \tag{2.2}$$

is related to the local density of states at the sample surface [*Hamers* (1992), *Feenstra* et al. (1987)]. The value of $\mathrm{d}I/\mathrm{d}V$ can be measured by analog or digital techniques, e.g., by modulation of V combined with lock-in detection of the change in I, or by measuring $I(V)$ curves at a fixed tip position. Several variations of this method are known as STS [*Feenstra* (1994)]. In this derivative mode, occupied as well as unoccupied states show up at the sample surface.

The effects of specific electronic states on STM images were discovered quite early [*Binnig* et al. (1983), *Becker* et al. (1985b), *Hamers* et al. (1986), *Feenstra* et al. (1987)]. They can be seen particularly well in semiconductors, where surface states in the gap produce sharp resonances. At metal surfaces the effects have been more subtle, which is understandable considering the continuous, free-electron-like density of states. Occasionally, special tip conditions have produced elemental contrast [*Chambliss* and *Chiang* (1992), *Schmid* et al. (1993), *Ruan* et al. (1993)]. However, there exists a general class of sharp surface states that are characteristic of metals. These are image states, i.e. electrons bound loosely by their image charge and thus interacting weakly with the metallic continuum [*Echenique* and *Pendry* (1990)]. Owing to their narrow widths, they produce large resonant tunneling effects and can be used to identify elements by their work function [*Jung* et al. (1995a)].

The dominant role of surface states in resonant tunneling is not surprising, given the long tails of their wave functions into the vacuum. However, there is a conceptual difficulty because surface states do not carry a current perpendicular to the surface [*Doyen* and *Drakova* (1993), *Drakova* and *Doyen* (1996)]. They are standing waves in that direction, whereas propagating waves are represented by bulk states. Several mechanisms can however lead to a transfer of electrons from surface to bulk states. First of all, surface states decay after a few femtoseconds into electron–hole pairs comprised mainly of bulk states. Also, the local nature of the STM tip breaks the symmetry of surface states and mixes them with bulk states [*Doyen* and *Drakova* (1993), *Drakova* and *Doyen* (1996)].

In addition to the resonant tunneling into surface states of semiconductors and metals there is "anti-resonant" tunneling into the band gap of insulators [*Viernow* et al. (1998)]. A topograph of a thin insulating film can be obtained using a sample bias large enough to allow tunneling into the conduction band of the insulator. The corresponding chemical map is obtained by acquiring a simultaneous current image at a bias voltage inside the gap of the insulator. Thereby, tunneling into insulating regions of the surface is suppressed compared to metallic and semiconducting regions.

2.2.2 Semiconductors: Dangling Bond States

A prototype for selective tunneling has been the Si(111)7×7 surface [*Binnig* et al. (1983), *Becker* et al. (1985b), *Hamers* et al. (1986), *Wolkow* and *Avouris* (1988)]. It exhibits two types of surface atoms with broken bonds (Fig. 2.3). The adatoms tie up three broken surface bonds and trade them for one broken bond; the rest atoms represent the remaining atoms of the truncated bulk structure. The orbitals are further stabilized by a transfer of electrons from adatoms to rest atoms, which gives rise to a lone pair at the rest atoms and nearly empty adatom bond orbitals. The respective orbitals can be seen with conventional spectroscopies, such as photoemission for occupied states and inverse photoemission for unoccupied states (Fig. 2.3b). They appear as

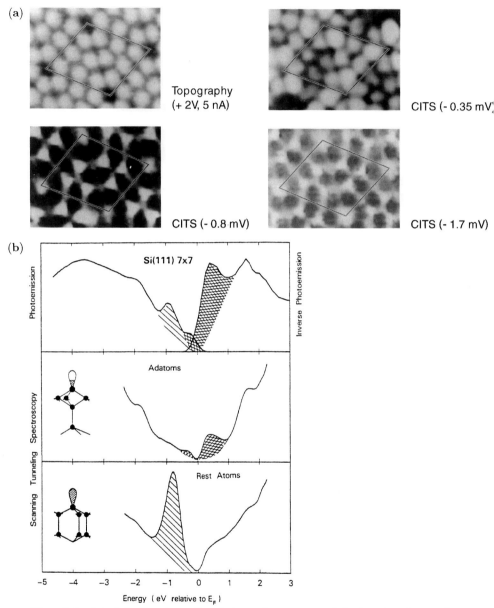

Fig. 2.3. (a) Selective tunneling into various types of broken bonds at the Si(111)7×7 surface. Broken-bond orbitals on rest atoms and adatoms show up in current images at fixed height for different bias voltages (−0.8 and −0.35 V, respectively), which correspond to their orbit energies. The tip height has been adjusted by a fixed sample bias of 2.0 V, 0.5 nA, at which there is little spectroscopic contrast. Adapted from [*Hamers* et al. (1986)], with kind permission of the authors. (b) Comparison of STS on specific sites [*Wolkow* and *Avouris* (1988)] with photoemission and inverse photoemission data at $k_\parallel = 0$. The rest atoms exhibit a completely filled, lone pair orbital, the adatoms a mostly empty broken-bond orbital. Reprinted from *Himpsel* (1990b), with kind permission of Elsevier Science – NL, Sara Burgerhartstraat 25, 1055 KV Amsterdam, The Netherlands

characteristic maxima in dI/dV tunneling spectra on top of adatoms and rest atoms (Fig. 2.3b, center and bottom). Taking current images at bias voltages corresponding to adatoms and rest atoms enhances them selectively (Fig. 2.3a). Rest atoms are barely visible in normal topographs.

Adsorbates on semiconductor surfaces display their own broken-bond states that distinguish them from the original surface atoms. To give a general guide to the electronic structure of adsorbates on silicon, we show photoemission and inverse photoemission spectra of group-III and group-V adsorbates in Fig. 2.4 [Himpsel (1990b)]. These are taken for zero parallel momentum $k_\parallel$ to the surface. This choice approximates STM where the parallel momentum distribution is centered around zero. The extra electron provided by As fills the broken-bond orbital and produces a stable lone pair. The missing electron in B produces an empty broken-bond state. In both cases, the unstable partially filled surface states have been removed from the gap. By tunneling into the filled state of a group-V adsorbate or into the empty state of a group-III adsorbate, it is possible to enhance specific adsorbate atoms. Examples of the latter are given in Figs. 2.5, 6 [Hamers and Demuth (1988), Avouris (1991), Chen et al. (1988)]; for a comprehensive review of STM of semiconductor overlayers see [Nogami (1994)]. Some care has to be taken when surface atoms interchange with adsorbates and produce a variety of geometries, such as B on Si [Lyo et al. (1989)], Si on B, and Si on Si in Fig. 2.6 [Avouris (1991)], cf. also P/Si [Bedrossian et al. (1989)], and halogens/Si [Pechman et al. (1995), Chander et al. (1995)]. There are many other geometries of how adsorbate atoms might be incorporated into semiconductor surfaces, e.g., dimers, trimers, and chains. Although their electronic states are more difficult to predict, it is quite common for them to have well-defined adsorbate states in or near the gap. By varying the coverage these states can be identified [Wang et al. (1994), P/Si(100)].

Cross-sectional STM proved capable of identifying atomic species in semiconductors using two different approaches. First, Salemink and Albrektsen (1993) identified Al atoms in cross sections of GaAs/AlGaAs multilayers on the basis of corrugation height differences in empty-state STM maps. Using this empty-state contrast, In atoms in InGaAs/GaAs superlattices can also be identified as shown in Fig. 2.7a, b. [Pfister et al. (1995)]. The In concentration in both the surface and the first subsurface layer can be obtained from a comparison of characteristic empty and filled-state images; see Fig. 2.7c.

In a second contrast mechanism, p-type dopant atoms have been recognized by means of a typical charge cloud with an approximate size of the extended Bohr radius. Be dopants have been characterized as far down as the fifth subsurface layer by Johnson et al. (1993); so have other dopants such as Si and Zn in GaAs. Further work has identified additional dopants in III–V semiconductors [n-type Si: Zheng et al. (1994a); p-type Zn: Zheng et al. (1994b)] and presented an atomic-scale analysis of dopant profiles [Be: Johnson et al. (1995)].

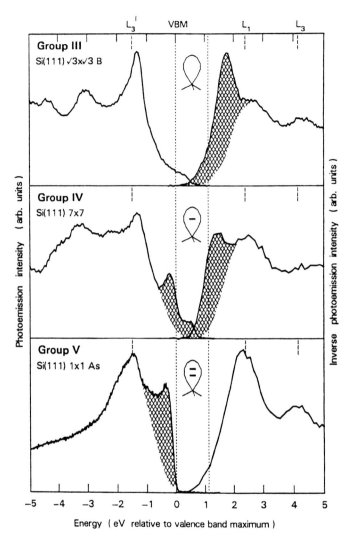

Fig. 2.4. Occupied and empty bond orbitals at silicon surfaces determined by photoemission and inverse photoemission at $k_\parallel = 0$. The clean Si(111)7×7 surface has partially filled broken-bond orbitals in the gap (center). These become filled upon adsorption of a group-V element (arsenic, *bottom*), and empty after adsorbing a group-III element (boron, *top*). From *Himpsel* (1990a)

As a consequence of the contrast mechanisms of semiconductors discussed above, Ballistic Electron Emission Microscopy (BEEM) [*Bell* and *Kaiser* (1988)] is also sensitive to atomic-scale variations in buried metal–semiconductor interfaces [*Sirringhaus* et al. (1995), *O'Shea* et al. (1997)]. Beyond surface features, buried defects can be imaged by "hot carrier scattering" [*Sirringhaus* et al. (1994)].

(a)

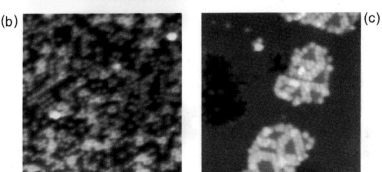

(b)

(c)

Fig. 2.5. STM image of Al on Si(111) showing Al and Si adatoms distinguished by different apparent heights [*Hamers* and *Demuth* (1988)]. Even though Al and Si are topographically similar at other bias voltages, they appear different when tunneling at positive (+2 V) sample bias, which emphasizes the broken-bond orbital of Si. At inverted bias, the Al atoms appear dominant. From *Hamers* (1989), with kind permission

Fig. 2.6. STM images of three different coverages of B on Si(111). (a) A image of the well-ordered B:$\sqrt{3}$ surface. Here B is observed to occupy a three-fold site underneath a Si adatom. Brighter $\sqrt{3}$ maxima in both filled and empty states are identified as substitutional defects with Si instead of B underneath the adatoms. (b) A B:$\sqrt{3}$ surface that is substantially B-deficient [30 % B saturation of the coverage shown in (a)]. Here characteristic features can be observed: The $\sqrt{3}$ arrangement of adatoms in two different heights due to the presence/absence of subsurface B and, additionally, a small density of ring-like features that are probably associated with excess Si. (c) A B:$\sqrt{3}$ surface prepared with excess B. The lower and higher areas have a different $\sqrt{3}$ structure, with depth and height smaller than the Si-bilayer step. This observation points towards a missing top layer structure. From *Nogami* (1994), Copyright 1994 World Scientific Publishing Company

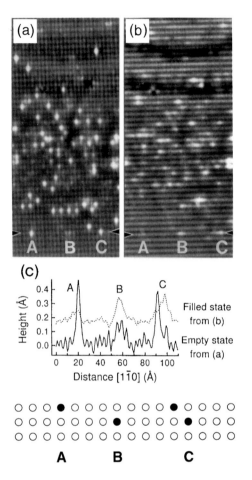

Fig. 2.7. Surface and subsurface recognition of In in $In_{0.12}Ga_{0.88}As/$ GaAs superlattices. (a) Empty and (b) filled-state STM images of identical cross sections (112×250 Å) correspond to the Ga/In and As sublattices, respectively. The In sites show up brightly in the empty-state image owing to geometric and electronic effects (larger covalent radius), whereas they geometrically distort the As positions, as imaged in the filled-state image. (c) Analysis of the corresponding features in both images leads to the identification of the surface In (A), subsurface In (B), and adjacent surface and subsurface In atoms (C). Adapted with permission from [*Pfister* et al. (1995)], Copyright 1995 American Institute of Physics

In metals the conduction electrons are much less localized than in semiconductors. Consequently, the detectable corrugation on metallic surfaces is typically lower (of the order of ~ 0.1 Å vs ~ 1 Å). Hence chemical contrast is more difficult to achieve, and therefore, alternative mechanisms have been developed.

2.2.3 Metals: Surface States and Image States

The electronic states that dominate tunneling at metal surfaces are usually the delocalized s, p-states, which extend farther away from the surface than the localized d-states. The s, p-like surface states can be classified in a rather simple scheme that starts with a series of image potential states [*Echenique* and *Pendry* (1990)], see Fig. 2.8. These states are due to electrons bound to the surface by their image charge and reflected back from the crystal by Bragg reflection in a band gap of bulk states. Even without a bulk gap the

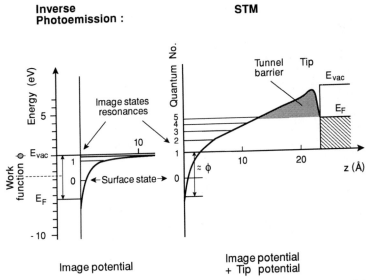

Fig. 2.8. Surface and image states in inverse photoemission and in STM. The extra electric field of the tip causes a triangular well potential that increases the characteristic energy of the image states and produces resonant states in the well. The position of the lowest image state ($n = 1$) is related to the work function ϕ, allowing the identification of metals. [Adapted from *Binnig* et al. (1985a), with kind permission of the authors]

wave function of the image state couples only weakly to the bulk continuum because it is located so far from the surface. A typical distance for the $n = 1$ state is 0.4 nm from the surface. The $1/4z$ image potential gives rise to a hydrogenic series of states with a binding energy of approximately 1/16 of a Rydberg below the vacuum level E_{vac} for the lowest state (principal quantum number $n = 1$). A somewhat more sophisticated approach includes the phase shift of the Bragg reflection and introduces a quantum defect d into the energy formula $E_n = E_{vac} - (\mathrm{Ry}/16)/(n + d)^2$. This allows extrapolation to $n = 0$. Such a state exists in many cases. It represents an intrinsic surface state or resonance, that is governed more strongly by the crystal potential than by the image potential. For example, the $n = 0$ state on Cu(111) and Au(111) is a p_z-like surface state split off from bulk states near the L_2 point. Similar p_z-like surface states exist on other low-index metal surfaces.

Surface and image states have been analyzed by inverse photoemission, which probes unoccupied states by letting an incident low-energy electron drop into an unoccupied state via emission of an ultraviolet photon. Figure 2.9 [*Mo* and *Himpsel* (1994)] shows inverse photoemission spectra for the $n = 1$ image state and the p_z-like $n = 0$ surface state for clean W(110) and for a Cu monolayer on W(110). With Cu on the surface, the image state shifts down and the surface state up. Although the shift of the $n = 0$ surface state is nontrivial, the shift of the $n = 1$ image state is easily understood: Image states

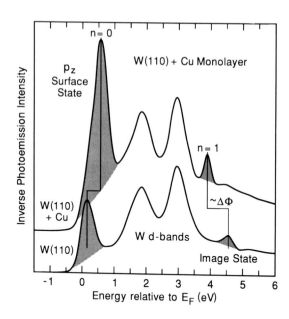

Fig. 2.9. Surface and image states on clean and Cu-covered W(110). The hatched states give rise to tunneling resonances and elemental contrast. A p_z-like, $n = 0$ surface state near the Fermi level and an $n = 1$ image state are seen in both cases. A similar situation is found for Cu/Mo(110) [*Jung* et al. (1995a)]. From *Himpsel* and *Ortega* (1994)

are referenced to the vacuum level, whereas inverse photoemission spectra and STM spectra are referenced to the Fermi level. The difference is the work function, and the shift from W to Cu simply reflects the work function change $\Delta\phi$ from W to Cu. Two-photon photoemission spectroscopy confirms analogous shifts of image-state series by the "local work function" of Ag on Pd(111) [*Fischer* et al. (1993a, 1993b)]. Surface states leave their mark in STS as well. For example, the p_z-like $n = 0$ surface state was used to display charge density ripples in STS of corrals and steps [*Crommie* et al. (1993b), *Eigler* (1994) on Cu(111); *Avouris* and *Lyo* (1994) on Au(111)]. It is interesting to note that the Cu-induced $n = 0$ state on W(110) in Fig. 2.9 is closely related to the Cu(111) state. The (111) surface of fcc Cu is close-packed. Likewise, the (110) surface of bcc W is the closest-packed bcc surface, which is simply a uniaxially elongated version of the hexagonally close-packed fcc(111) surface. As a Cu monolayer grows in registry with W(110), the topology of the bands is comparable. STS on metals resolves quantum interference effects of field emission resonances, as can be learned from numerical model calculations [*Kalotas* et al. (1996), *Pitarke* et al. (1990), *Gundlach* (1966)].

Chemical contrast has been obtained from STM on metals via surface and image states. Figure 2.10 demonstrates the effects observed using a surface with alternating stripes of Cu and Mo separated by steps. Cu stripes were produced by step-flow growth on a vicinal Mo(110) surface misoriented by 0.2°. As the first Cu layer is in registry with Mo there is practically no topographical contrast between the Cu stripes and the upper Mo terraces. Chemical contrast is important here to determine the growth mode [*Himpsel* et al. (1994), *Jung* et al. (1995b)]. For most bias voltages there is little dis-

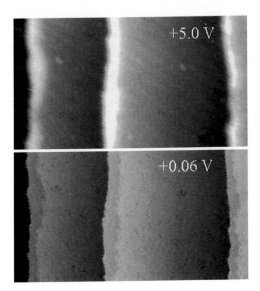

Fig. 2.10. Distinguishing Cu from Mo via tunneling into a Cu image-state (top, Cu stripes bright) or a Mo surface state (bottom, Mo stripes bright). When tunneling through the $n = 0$ surface resonance of Mo(110) at +0.06 V sample bias (bottom), nearly atomic resolution is obtained (note the black dots corresponding to single adsorbed oxygen atoms). Using the $n = 1$ image state resonance of Cu at +5 V sample bias, the picture is blurred by about 1 nm. Adapted from *Jung* et al. (1995a)

tinction between Cu and Mo. However, by resonant tunneling into the $n = 0$ surface state of Mo (bottom) or the $n = 1$ image state of Cu (top) it is possible to make Mo or Cu appear brighter in a constant-current image. Likewise, enhancement of Cu due to its $n = 0$ surface has been found for Cu on W(110) and Mo(110) [*Mo* and *Himpsel* (1994), *Himpsel* et al. (1994), *Jung* et al. (1995b)], Cu appears brighter than Mo and W when the sample is biased close to the Cu resonance energy of 0.6 and 0.8 eV, respectively (Fig. 2.10).

Image states provide higher elemental contrast than do surface states. They are narrower, owing to their longer lifetime [*Steinmann* and *Fauster* (1995)], and the weaker damping leads to a higher resonant amplitude. Figures 2.11–13 characterize resonant tunneling through image states for Cu on Mo(110) [*Jung* et al. (1995a, 1995b)]. At normal tunneling voltages the Cu stripes cannot be discerned (Fig. 2.11, center). Adjusting the sample bias to the $n = 1$ image-state resonance makes the Cu stripes appear bright (Fig. 2.11, top). The difference picture (Fig. 2.11, bottom) subtracts out the topography and leaves us with an elemental map of the surface. To locate the image-state resonance more precisely, the sample bias is varied continuously from top to bottom in Fig. 2.12. Only a narrow region of a few tenths of 1 eV around the $n = 1$ resonance exhibits the resonant enhancement. Figure 2.13 quantifies the resonant enhancement and extends the energy range to higher resonances ($n = 2, 3$). At resonance the elemental contrast can reach a full step height in the constant-current mode, i.e., about 0.2 nm. With a typical tunnel current increase of one order of magnitude per 0.1 nm, this is an enormous effect. The energy position of image states is somewhat shifted in STM owing to the high bias voltages that

Fig. 2.11. Chemical contrast by resonant tunneling. *Top*: Resonant tunneling via the $n = 1$ image state of Cu enhances tunneling into Cu stripes that are grown in the step-flow mode on Mo(110). *Middle*: Nonresonant tunneling yields the topography. *Bottom*: The difference eliminates the topography and provides a chemical map of the surface. From *Himpsel* et al. (1994)

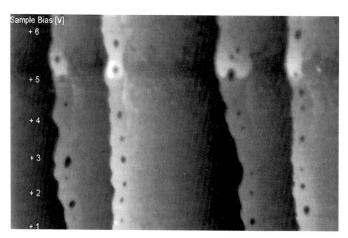

Fig. 2.12. Resonant tunneling via the $n = 1$ image state of Cu stripes on stepped Mo(110). The sample voltage was increased from the bottom to the top of the constant-current image while scanning. Chemical contrast is maximum at bias voltages within a few tenths of an eV around the resonance. The Cu stripes are marked by dark holes caused by adsorbed oxygen. From *Himpsel* et al. (1996)

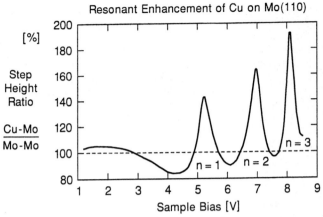

Fig. 2.13. Quantification of the chemical contrast versus sample bias for the Cu stripes on Mo(110) in Fig. 2.12. The first three image resonances are seen as an apparent increase in the Mo to Cu step height. A large enhancement is observed because of the narrow tunneling resonances. Reprinted by permission from *Himpsel* et al. (1998b), Copyright 1997 by International Business Machines Corporation

are needed to reach them ($>5\,\text{V}$). The electrostatic potential induced by the tip modifies the image potential and pushes the states upward (Fig. 2.8) [*Binnig* et al. (1985a), see also *Becker* et al. (1985a), *Coombs* and *Gimzewski* (1988)]. These image-state resonances in STM have been described by *Kalotas* et al. (1996), using a model with variable STM tip radius and inclusion of the contribution from the image potential to the barrier. Therefore, the $n = 1$ image state is seen as resonance at about $+5\,\text{V}$ bias in Cu, depending on the tip radius, and the higher states are pushed up even further. This upward shift affects Cu and Mo states alike, such that the difference in the image-state positions seen by STM is still a good measure of the work function difference. Work functions are tabulated for many metal surfaces and adsorbate systems [*Hölzl* and *Schulte* (1979)] and thus make it easy to identify surface species.

Instead of using image-state resonances to track the local work function, it is also possible to measure the work function by determining the tunneling barrier ΔE [Fig. 2.2, (2.1)], which is somewhat smaller than the work function. This can be achieved by modulating the distance z instead of the sample voltage. From (2.2) we find that

$$(\Delta E + \text{eV} - E) \sim \left[\frac{\text{d}(\ln I)}{\text{d}z}\right]^2 = \left[\frac{\text{d}I}{\text{d}z\,I}\right]^2 , \qquad (2.3)$$

i.e., a $\text{d}I/\text{d}z$ map represents the local work function. This method has been used to distinguish Fe stripes on stepped W(110) by their lower work function [*Wiesendanger* et al. (1996)]. Using STM the thermovoltage, which is related to the electronic states of a surface, can also be mapped. Here both atomic contrast [*Williams* and *Wickramasinghe* (1990)] and the identification of Cu islands on a Ag(111) surface [*Rettenberger* et al. (1995)] have been demonstrated (Chap. 3).

In addition to the delocalized s, p-like surface and image states, there are d-like surface states that might, in principle, be useful for resonant tunneling and chemical contrast. As the $3d$ states in the magnetic transition metals Cr, Mn, Fe, Co, and Ni are spin-polarized, it is conceivable to utilize spin-dependent tunneling to map magnetic domains at surfaces (Chap. 4). However, $3d$ states are more localized than $4s, p$ states, leading to a faster decay of the wave function into the vacuum and less overlap with the tip states. A signature of their shorter decay has already been obtained by measuring the spin polarization of the tunneling current at various tip–sample distances [*Alvarado* (1995)]. It increases towards shorter distances as the strongly polarized d-states take over from the weakly polarized s, p-states. In noble metals the d-states move down in energy, become even more localized, and are invisible to STM. Despite the fast decay, a surface state with d-character has been identified on Fe(100) and Cr(100) surfaces that appear in dI/dV spectra [Fig. 2.14, *Stroscio* et al. (1995)]. *Davies* et al. (1996) have been using the attenuation of this surface state due to Cr adsorbates to identify isolated Cr atoms on a Fe(100) surface. Fe-induced surface states have also been seen on

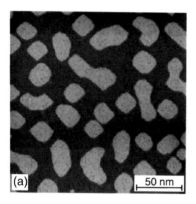

Fig. 2.14a, b. Atomic-scale contrast of Cr on Fe: (a) Cr island ($\sim$ 0.4 ML) grown on Fe(001) at a sample bias of -1.1 V. (b) High-resolution STM image of the Cr-exposed substrate showing alloyed impurity atoms. STS experiments reveal attenuation of the Fe surface state in the vicinity of the impurity atoms and allow the identification of the Cr impurities. The inset shows simulated pairs of Cr impurities separated by first, second, third and fourth nearest-neighbor (nn) distances. There is no occurrence of first nn Cr pairs in the experimental data as a consequence of the alloying mechanism. Adapted from *Davies* et al. (1996), with kind permission of the authors

W(110), but their character has not yet been established [*Wiesendanger* et al. (1996), *Bode* et al. (1996a, 1996b)]. For an Fe(100) surface alloyed with Si, a one-dimensional surface state localized along the Fe chains has been characterized using STS [*Biedermann* et al. (1996)]. The assignment of this surface state has been performed based on first-principles spin polarized calculations, as the $3d$ states map magnetic domains. A detailed review of complementary experiments regarding electronic states in magnetic nanostructures can be found in *Himpsel* et al. (1998a, 1998b).

2.2.4 Influence of Tip States

Resonant tunneling depends not only on surface states but also on tip states. Extreme cases to consider are either a continuum of tip states interrupted by the Fermi level or a discrete level at the tip. So far we have considered the first case, which is typical for metallic tips with a continuous spectrum of states. A sharp tip level, on the other hand, indicates an atomic state of a loosely bound tip atom. Such a situation tends to be unstable and is the exception rather than the rule. If it could be controlled reproducibly, such a tip would yield stronger tunneling resonances than the step function induced by the Fermi level cutoff. Of course, the energy of the tip level has to be known because it affects the bias voltage where the tip level and a sample level coincide.

On semiconductors, tip-dependent variations in tunneling spectra have been observed and modeled [Fig. 2.15, *Avouris* (1991)]. In some instances, a

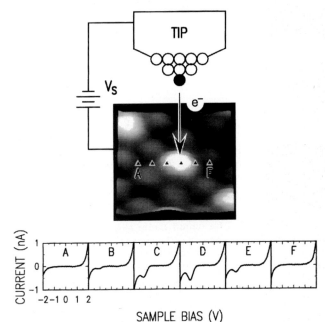

Fig. 2.15. Topographic image and tunneling spectra across an adatom on a B-covered Si(111) surface. Tunneling between discrete states at the sample and the tip gives rise to a maximum in the $I(V)$ curve and negative differential resistance dI/dV. From *Avouris* (1991), with kind permission of the author

combination of a sharp defect state at the sample surface and a sharp tip state was found. When these two levels are scanned across each other by varying the sample bias they produce a peak in the $I(V)$ spectrum at the bias where they coincide. Consequently, the differential resistance dI/dV will turn negative on the high bias side of the resonance.

On metals, special tip conditions can give rise to chemical selectivity. For example, Cu and Au patches could be distinguished by an adsorbate on the tip [*Chambliss* and *Chiang* (1992)]. Using a W tip with or without an oxygen atom adsorbed, it was possible to image oxygen or nickel atoms selectively at an oxygen-covered Ni(110) surface [*Ruan* et al. (1993)]. In Pt–Ni alloys it was possible to discern individual Ni and Pt atoms under special tip conditions, which yielded an apparent height difference of 0.3 Å [Fig. 2.16a, *Schmid* et al. (1993)]. The relevant electronic states have not yet been assigned in these cases. For resonant tunneling at Cu/Mo(110) the surface states of the sample are known from inverse photoemission experiments (see Fig. 2.9). In Fig. 2.17, the $n = 0$ image state is used to enhance tunneling into Cu. In

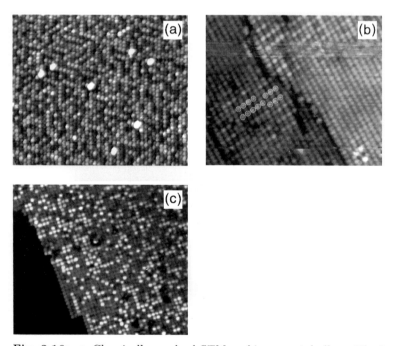

Fig. 2.16a–c. Chemically resolved STM on binary metal alloys. The images correspond to a scanned area of 12.5 nm×10 nm. (a) $Pt_{25}Ni_{75}$ (111) imaged by STM (5 mV, 16 nA). Special tip conditions, possibly associated with an adsorbate, allow the identification of Pt (dark) and Ni (bright) atoms. (b) $Pt_{10}Ni_{90}$ (110) imaged by STM. Here a 2×1 superstructure of the top monolayer can be observed (c) $Pt_{0.5}Rh_{0.5}$ (100) imaged at 0.5 nA and 0.5 mV. Individual Pt (dark) and Rh (bright) atoms can be identified. Adapted from (**a**): *Schmid* et al. (1993); (**b**): *Ritz* et al. (1996), and (**c**): *Wouda* et al. (1996). With kind permission of the authors

Fig. 2.17. Tip dependence of resonant tunneling through the $n = 0$, p_z-like surface state of Cu stripes on Mo(110). Changes in the tip during the scan modify the chemical sensitivity. Sample bias is 0.9 V

this particular picture, the tip was unstable and jumped back and forth between two geometries, one with high-resonant enhancement of Cu, the other without a resonance. Although such an instability is atypical for this surface, it demonstrates how much the tip may affect resonant tunneling near the Fermi level such an influence of tip shape and the resulting electronic structure has recently been observed in spectroscopy performed with W tips on the Cu(111) substrate [*Vázquez de Parga* et al. (1998)]. Image-state resonances at high bias voltages are less dependent on the tip condition. Because of their different origin, there are always sharp tunneling states above metallic surfaces. For example, the resonance curves in Figs. 2.12, 13 were obtained with a tip that exhibited little contrast when tunneling through the $n = 0$ surface state. Thus, image states provide a reliable source of elemental contrast for metals. Whereas image states are ubiquitous on metals, they have not been found on semiconductors, probably due to the high density of bulk states near the vacuum level. Tip-induced resonances are observable by STM on semiconductors [*Binnig* et al. (1985a), *Becker* et al. (1985a)], but not as sharply as on metals. Thus, an equivalent mechanism for chemical identification on semiconductor surfaces has yet to be found.

2.2.5 Lateral Resolution

In order to find the limits for the spatial resolution of resonant tunneling we have to consider electronic states at steps and small patches of atoms. If they are different from the states of an extended surface, they will produce different resonance conditions. In a sense, changes of the electronic states at the edge of a patch reflect altered chemical characteristics of those atoms. The broken-bond states at semiconductor surfaces are rather directional and localized, such that atomic resolution is possible for chemical analysis. Metal states spread out and tend to blur the chemical STM map by up to 1 nm.

Our model case of resonant tunneling via surface and image states in Cu/Mo(110) demonstrates how the resolution of the chemical map depends on the resonant state. In Fig. 2.10 two different states are used for chemical contrast, i.e., the $n = 0$ surface state of Mo(110) just above the Fermi level and the $n = 0$ image state of Cu at +5 eV. Tunneling close to the Fermi level produces the sharper picture. In particular, narrow parts of the Cu stripes are clearly continuous in the image at +0.06 V bias but appear pinched off for +5 V bias. It is not surprising that the resolution is degraded at high bias voltages because the tip–sample distance is increased to keep the tunnel current constant. However, there is a possible additional broadening due to the finite lateral extent of the image state itself.

The lateral extent of surface states at step edges has been mapped by STS [*Crommie* et al. (1993b), *Eigler* (1994), *Avouris* and *Lyo* (1994)]. Figure 2.18 gives an example for which the charge density oscillations of the $n = 0$ surface state of Cu(111) are mapped out at a step edge. Within about 1 nm of such an edge, the regular oscillations expected from the $E(k)$ band dispersion of an infinite, two-dimensional state are distorted. Another stepedge effect is driven by structural changes. The fringes around the edges of Fe islands on W(110) in Fig. 2.19 have been assigned by *Wiesendanger* et al. (1996) and *Bode* et al. (1996a, 1996b) to a partial strain relaxation. Again, they are about 1 nm wide. A more abrupt change of the electronic structure has been inferred from inverse photoemission for a single string of Cu atoms bound at a W(110) step edge. Compared to additional Cu rows, the energy of the $n = 0$ surface state of the first Cu row is lowered by 0.4 eV owing to the influence of the neighboring W atoms [*Himpsel* and *Ortega* (1994)].

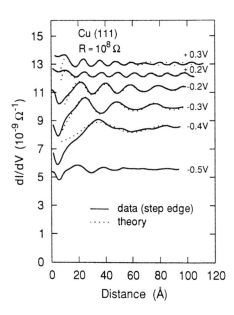

Fig. 2.18. Spatial modulation of the $n = 0$ p_z-like surface state on Cu(111) at a step edge. Depending on the bias voltage the charge density oscillates with varying wavelength, as expected from the $E(k_\parallel)$ band dispersion of the surface state. Within about 1 nm of the step edge there are deviations from the regular oscillations [*Crommie* et al. (1993b)]. From *Eigler* (1994), with kind permission of the author

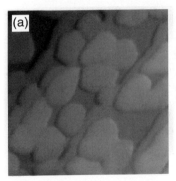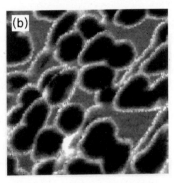

Fig. 2.19a, b. Spectroscopic changes at the edge of Fe islands on W(110). Compared to the constant-current topography islands appear wider or narrower by 1–2 nm in spectroscopic dI/dV images, depending on the sample bias. This is explained by a change in the electronic structure at the step edge triggered by strain relaxation. See also *Bode* et al. (1996a), *Bode* et al. (1996b). (a) Adapted from *Bode* et al. (1996b), (b) from *Wiesendanger* et al. (1996), with kind permission of the authors

The ultimate, atomic resolution limit can be tested by looking at the extent of single surface atoms by STM. Figure 2.20 shows the image of a single Fe atom on Pt(111), which is rather wide [7 Å full width at half maximum, *Crommie* et al. (1993a)]. This large width reduces the corrugation observed for a surface and the chemical contrast that can be achieved. Adsorbate-induced special tip contrast [*Ruan* et al. (1993), *Schmid* et al. (1993)] has proven capable of mapping the composition on an atomic scale (Fig. 2.16a). Comparable contrast has only been achieved for $Pt_{10}Ni_{90}(110)$ (Fig. 2.16b)

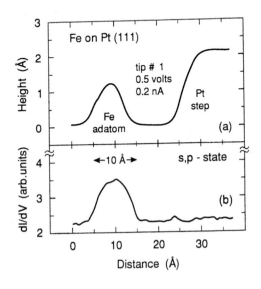

Fig. 2.20. STM image of a single Fe atom on Pt(111), giving a full width at half maximum of 10 Å. From *Crommie* et al. (1993a), with kind permission of the authors

and $Pt_{0.5}Rh_{0.5}$ (Fig. 2.16c) as a consequence of tip-independent variations of the density of states at the Fermi level (Schmid, personal communication).

The d-like, one-dimensional surface state localized along Fe chains in an Fe–Si alloy, as observed by *Biedermann* et al. (1996), has a very pronounced localization. This specific surface state allows nearly atomic resolution of elemental contrast in one dimension. These examples of high-resolution atomic-composition imaging via Density Of States (DOS) variations or special tip states strongly depend on the physics and chemistry of sample and tip. In contrast, the image-state mechanism provides a more universal scheme, in which atomic information can be supplemented by complementary experiments.

2.3 Probing Atomic and Molecular Properties

The preceding section dealt with chemical recognition via spectroscopies of electronic states using STM. Non- or poorly conductive surfaces lacking pronounced electronic states require alternative scanning probe microscopies (AFM, SNOM) and contrast mechanisms to detect interfacial chemical composition. Atomic and molecular resolution contrast in SPM allows the analysis of structural issues [for an overview, see *Magonov* and *Whangbo* (1996)]. The recent progress in contact [*Howald* et al. (1994)] and true noncontact force microscopy [*Giessibl* (1995), *Sugawara* et al. (1995)] may trigger further improvements regarding image quality, especially on soft molecular materials. Atomic or molecular resolution micrographs indirectly provide composition information. For example, the analysis of atomic radii [*Chambliss* and *Chiang* (1992)] can resolve the atomic composition of a surface layer. Modeling a molecular structure and imaging [*Gould* et al. (1988), *Meyer* et al. (1991), *Overney* et al. (1992a, 1992b)] can provide information about the orientation of molecules at the surface of a molecular crystal. This is restricted to simple cases, however, as the differences in the Van der Waals (VdW) radii of most elements are small and close to instrumental resolution limits ($\lesssim 0.1\,\text{Å}$). Some surfaces exhibit surface reconstructions [*Woodruff* (1994), *Wintterlin* and *Behm* (1992)], where the atomic arrangement depends critically on atomic composition and/or adsorbates.

The selective adsorption of a reactive gas (O_2) allowed the identification of Cu patches after adsorption of Cu onto a Co substrate [*Allenspach* and *Dürig* (1996), *Allenspach* et al. (1997)]. For example, the multistage adsorption and decay of $Sb_4 \rightarrow Sb_2$ on the Si(100) dimer rows has been unraveled in detail [*Mo* (1992)]. Recognition of atomic or molecular adsorbates [*Mo* (1993a)] in time-lapsed imaging sequences enables the determination of diffusion barriers [*Mo* (1993b)]. More complex reactions have been monitored using SPM techniques: The catalytic decay of NO on Ru(0001) has been traced [*Zambelli* et al. (1996)], and intermediate adsorption/reaction stages were resolved in the reaction of oxygen (O_2) with $TiO_2(110)$ at 800 K [*Onishi* and *Iwasawa* (1996)].

2.3.1 Molecular Shape in Scanning Probe Microscopy

As the preparation techniques for molecular layers have become increasingly sophisticated, numerous reviews and books have appeared [*Frommer* (1992), *Fuchs* (1993), *Morozov* et al. (1993), *Magonov* and *Whangbo* (1996)]. Recognition of the top-layer molecular species enables one to determine molecular packing and orientation through series of scanning probe experiments and comparison with theoretical models [*Sautet* and *Joachim* (1991, 1992), *Chavy* et al. (1993)]. For example, *Hallmark* et al. (1991), *Chiang* (1992) and *Hallmark* et al. (1993) have demonstrated the identification of naphthalene-derived polycyclic aromatic compounds adsorbed on metals using STM and extended Hueckel-type calculations. Similar contrast has recently been demonstrated for different thiophenes [*Frank* et al. (1995)]. Figure 2.21 presents another example of two porphyrin compounds that differ slightly in the length of their substituents and have been coadsorbed onto a metal substrate in ultrahigh vacuum. The individual molecular units can be recognized based on the different shape of the substituents and the different contrast. For the di-tb-phenyl-substituted porphyrin, STM contrast is dominated by tunneling through the substituents. The porphyrin group is imaged on the substrate level, and thus does not contribute significantly to the tunneling path. In the case of the tetraphenyl porphyrin, X(010) and X(001), the porphyrin center unit as well as its substituents are imaged as cross-shaped depressions using tunneling parameters of 100 pA and 2000 mV. Thus the tunneling contribution of the molecule's electronic states has characteristically changed between the two variations of the porphyrin compound.

In addition to the contrast due to varying substituents, *Hipps* et al. (1996) and *Lu* et al. (1996) have demonstrated the recognition of different center metal atoms in a mixture of Cu-phthalocyanine and Co-phthalocyanine. Here a d-orbital contribution to the tunneling close to the Fermi level has been proposed as the essential contrast mechanism in agreement with theoretical calculations. Mixed porphyrin adlayers have also been characterized at the solid–liquid interface using potential-tuned resonant tunneling [*Tao* (1996)]. In this case, the contrast is believed to depend on the different redox properties, as STM detects the variances in electron transfer between substrate and adsorbed molecules resulting from an applied bias voltage sweep. These results clearly demonstrate the capability of SPM to identify individual molecular units based on real-space maps of surfaces and interfaces [*Jung* et al. (1997b)]. In close-packed molecular layers of Self-Assembled Monolayers (SAMs) of thiols on gold [*Delamarche* et al. (1996)], STM revealed contrast on individual tail groups of mixed layers [*Takami* et al. (1995)]. The orientation of the tails close to the domain boundaries [*Schönenberger* et al. (1995)] and molecular order have also been resolved in STM experiments

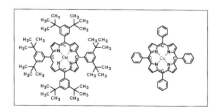

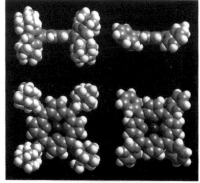

Fig. 2.21. Molecular identification of Cu-tetra [di-tb-phenyl] porphyrin (B) and tetraphenyl porphyrin X(010) coadsorbed onto a Cu(100) substrate. High-resolution STM maps allow identification of the molecular species based on inverted contrast and recognition of individual substituents. For the larger porphyrin there exists only one adsorption state, B on Cu(100), whereas the phenyl substituents in tetraphenyl porphyrin are antisymmetrically rotated, leading to two adsorption states: X(010) and X(001). Thus not only the identity of the adsorbed species but also its conformation are accessible from these experimental data [*Jung* et al. (1997a)]

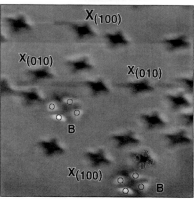

[*Wolf* et al. (1995), *Sprik* et al. (1994)]. In conclusion, both STM and AFM allow the identification of molecular units based on the analysis of molecular shape in experiment and simulation. Similar to the case of metals, current–voltage spectroscopy allows complementary contrast in many cases. Analogous modes in force spectroscopy exist and will now be discussed in terms of their relevance to chemical information.

2.3.2 Lateral Force Microscopy and Elastic Properties

Similar to the case with STM, AFM images have primarily been interpreted as topographical maps of substrates. Depending on the choice of force sensor, however, different types of interaction forces, electrostatic, attractive (VdW), magnetic, or repulsive forces can be mapped. Furthermore the magnitude of the force interaction is selectable in a range of 10^{-6} to 10^{-9} N for contact force microscopy and down to 10^{-13} N for noncontact force microscopy [*Meyer* and *Heinzelmann* (1992)]. Two additional experimental modes are important in the context of identifying chemical species: The simultaneous detection of lateral forces [*Marti* et al. (1990), *Meyer* and *Amer* (1990)] and the simultaneous measurement of applied force and sample deformation in a force vs distance curve [for a collection of papers on forces in SPM, see *Güntherodt* et al. (1995)].

Both perpendicular and lateral forces and the elastic/plastic response can be utilized to derive information related to the chemical composition of surfaces. Force vs distance curves have been used to map adhesion differences on patterned surfaces. The quantitative analysis of adhesion and frictional force at given load allows the identification of materials: At the solid–liquid interface, *van der Werf* et al. (1994) have identified a polymerized Langmuir–Blodgett film and a mercaptopentadecane-gold layer on glass substrates. Domains of mixed Cd-arachidate layers with differently substituted end groups have been identified [*Meyer* et al. (1992)]. Figure 2.22 illustrates how patches of different Cd-arachidates have been arranged into the shape of "smileys" using an AFM to rearrange the domains. The hydrocarbon and fluorocarbon chains can be differentiated in a frictional force image. Using cantilevers sensitized with well-defined end groups by a SAM technique, *Frisbie* et al. (1994) have identified micropatterned patches with COOH and CH$_3$ end groups. Lateral forces have also been measured at the liquid–solid interface

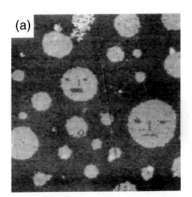
(a)

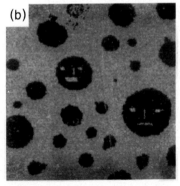
(b)

Fig. 2.22a, b. Force microscopy image $(2.4 \times 2.4 \,\mu\mathrm{m}^2)$ of a 1:1 molar mixture of arachidic acid (C$_{19}$H$_{39}$COOH) and partially fluorinated ether carboxylic acid (C$_9$F$_{19}$C$_2$H$_4$–O–C$_2$H$_4$COOH). After deposition by Langmuir–Blodgett technique [*Roberts* (1990)], circular domains of the two compounds can be identified by the topographic contrast mapped in (a) (characteristic length of the compounds). The "smiley" pattern has been generated by force microscopy modification. The friction force map shown in (b) identifies two different friction patterns for hydrocarbons as opposed to fluorocarbons. Reprinted from *Meyer* et al. (1992), with kind permission of Elsevier Science S.A., P.O. Box 564, 1001 Lausanne, Switzerland

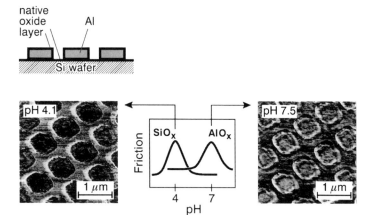

Fig. 2.23. pH-dependent lateral forces used to distinguish inorganic surfaces [*Marti* et al. (1996a, 1996b), *Hähner* et al. (1997)]. SiO_2 substrate patterned on the submicrometer scale with AlO_x studs. Friction maps of a surface area of $3\times3\,\mu m^2$ identify different peak friction values at pH 7.5 for the AlO_x (*left*) and pH 4.1 for SiO_2 (*right*). The general dependence of the measured differential friction forces on pH is indicated in the center. Adapted from *Hähner* et al. (1997), with kind permission of the authors

[*Marti* et al. (1995)]. In a recent experiment the same group proved pH-dependent frictional contrast of two inorganic compounds, SiO_2 and AlO_x [*Marti* et al. (1996b), *Hähner* et al. (1997)], as shown in Fig. 2.23. Similar experiments have been used to identify the components of a polymer blend [*Feldman* et al. (1997)]. Another important technique to identify monolayer materials is the local measurement of elastic properties using modulated forces perpendicular to the sample surface [*Overney* et al. (1994)]. These techniques have proven to be very suitable for identifying well-defined mixtures of monolayers applied to the surface by either epitaxial growth, Langmuir–Blodgett or self-assembly techniques [see *Roberts* (1990), *Ulman* (1991), *Delamarche* et al. (1996), and references therein]. SAM films with varying substituents can be structured into predefined patterns using microcontact printing techniques [*Kumar* and *Whitesides* (1993), *Kumar* et al. (1994), *Biebuyck* et al. (1997), *Wilbur* et al. (1994)]. Similar to the Langmuir–Blodgett system, SPM techniques applied to SAMs allow versatile contrast [*Wilbur* et al. (1995)] to be obtained both from the detection of lateral forces and from modulated force microscopy. Contrast depends critically on the nature of the end group either by chemical (shell-electron property of outermost atoms) interaction with the sensor or by changes in the contact area due to the different molecular elasticity. The latter is evidenced by a recent study of frictional forces on mixed chain length, phase-segregated Langmuir–Blodgett monolayers where frictional contrast is dominated by other factors than the chemistry of the end group exposed to the force sensor [*Barger* et al. (1996)]. Figure 2.24 demonstrates the observed predominance of mechanical contrast

Mechanical Contrast

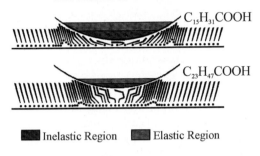

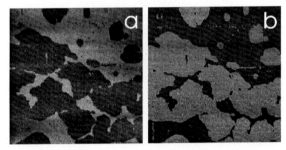

Fig. 2.24. Mechanical contrast in dependence of the chemical composition on mixed Langmuir–Blodgett films with two chain lengths (C_{16} and C_{24} fatty acids). *Upper image*: schematic of the compression contrast mechanism. *Lower images*: maps showing the topography (**a**) and adhesion (**b**) deduced from a force curve array. LB film domains with different molecular compositions can be distinguished. Reprinted from *Koleske* et al. (1997), with kind permission of the authors. This mechanical contrast mechanism is noticeable in adhesion or friction force measurements [*Koleske* et al. (1997)], and may also be relevant to STM imaging, where forces of the order of 10^{-9} N have been observed [*Dürig* et al. (1990)], as well as affect tunneling through the deformation of molecular units

over the contrast of chemical forces between the sensor and the contacting end groups [*Koleske* et al. (1997)]. Similar to many of the previously described examples, this allows separation of two phases. The single parameter available, however – here the frictional force – is a convolution of a few physical and chemical properties that are related to the structure of the chemical compound rather than a simple function of the generally complex composition of the interfacial layer.

2.3.3 Optical and Spectroscopic Techniques

Similar to the local detection of mechanical properties, the characterization of electromagnetic properties in a SPM junction gives access to information that relates to chemical information and complements the high-resolution topographs. This approach offers several attractive features. First, photons represent a particularly versatile channel of information. Their intensity, spectral distribution, angle of emission or incidence, polarization status, and time correlation are accessible and represent unique probes for chemical identification. Spatial mapping of these physical quantities permits the addition of "true color" to SPM images. Second, the emission or absorption of electromagnetic radiation is a chemically specific characteristic of many excitations of solids and molecules. This encompasses phenomena such as plasmons and interband transitions in metals, intrinsic luminescence and luminescent defects in semiconductors, molecular fluorescence, and modified optical properties owing to quantum size effects in metal and semiconductor particles and nanostructures.

Various papers by Pohl and coworkers [*Pohl* et al. (1984, 1985), *Dürig* et al. (1986)] and by Lewis, Betzig and Trautman [*Betzig* et al. (1986, 1987), *Betzig* and *Trautman* (1992)] demonstrated the resolution capabilities and limits of SNOM. For reviews, the reader is referred to *Pohl* (1991, 1992), *Betzig* and *Trautman* (1992), *Heinzelmann* and *Pohl* (1994). Although undoubted optical resolution is of the order of 30 nm, higher resolution has been claimed [*Zenhausern* et al. (1995)]. In spite of the ambiguous resolution limits, dilute single molecules with characteristic chromophores have been imaged [*Betzig* and *Chichester* (1993)], and analyzed using fluorescence characteristics [*Trautman* et al. (1994)]. *Paesler* and *Moyer* (1996) give a general introduction into SNOM-based techniques. In combination with polarized light, SNOM has been used to alter and determine surface magnetization with the magneto-optic Faraday effects. Similarly the crystal structure of liquid crystals has been altered and subsequently analyzed on the submicrometer scale [*Moyer* et al. (1995)]. In a comparable size range, polarization fluorescence has been applied to identify UV polarized grains of diethylene glycol diamine pentacosadiynoic amide with different orientations [*Moers* et al. (1994)]. As the typical resolution of most SNOM-based experiments remains in the submicrometer range, chemical imaging will require labeling with one or the other adequate chromophores.

An alternative approach to optical spectroscopy and mapping is to use the localized filament current of the STM tip to generate photon emission locally. The first experimental evidence of STM-induced light emission from metals and semiconductors was published in 1988 [*Gimzewski* et al. (1988)]. The photon emission in metals is attributed to the radiative decay of localized, tip-induced plasmon modes which are excited predominantly by inelastic tunneling processes [*Johansson* et al. (1990), *Berndt* et al. (1991), *Persson* and *Baratoff* (1992), *Uehara* et al. (1992), *Berndt* et al. (1993b)].

Surface emissivity patterns have been proved to map atomic [*Berndt* et al. (1995)] and molecular patterns [*Berndt* et al. (1993a)]. Selectivity has been demonstrated for nanometer-sized clusters of W on Cu(111), where W was found to quench the photon emissivity of the Cu substrate strongly. Another example is shown in Fig. 2.25 [*Berndt* et al. (1991), *Berndt* and *Gimzewski* (1993), *Gimzewski* (1995)]. Here the photon emissivity of TiO_2 clusters on

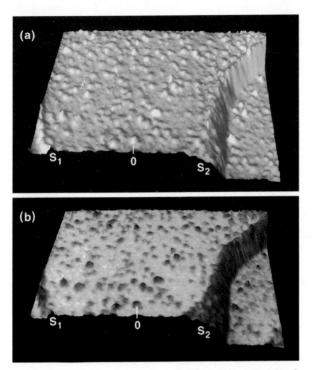

Fig. 2.25a, b. Discrimination of chemical composition based on a photon emissivity map and STM topograph: **(a)** STM topograph of a $1000\,\text{Å} \times 1000\,\text{Å}$ area of a Ti film that has been exposed to $20\,\text{L}$ of O_2 at room temperature. **(b)** Photon map recorded simultaneously with the topographic image, shown in **(a)**. Although the topographic height of the features is about $2\,\text{nm}$, the photon intensity drops from ~ 1000 to ~ 20 counts per second. Reprinted from *Berndt* et al. (1992), with kind permission of Elsevier Science – NL, Sara Burgerhartstraat 25, 1055 KV Amsterdam, The Netherlands

a Ti substrate is observed to be significantly lowered. Similarly, selective chemoluminescence has been applied to map CdS and GaAs [*Berndt* and *Gimzewski* (1992)].

An alternative way to perform molecular spectroscopy in STM through inelastic tunneling processes has been proposed by *Binnig* et al. (1985b). With the recent achievements in STM instrumentation, it seems plausible that Inelastic Tunneling Spectroscopy (IETS) experiments performed in the STM junction can contribute to IR and Raman data in the vibrational regime [*Hipps* and *Mazur* (1993)].

The inverse process, i.e., the observation of changes in tunneling parameters upon irradiation of the junction with light, has also been used to gather information: *Kazmerski* (1991) detected atomic composition on semiconductor surfaces, whereas *Gimzewski* et al. (1991) provide experimental evidence that photon electron emission characteristics can be probed locally using

STM. In the context of SPM-derived sensing of local electromagnetic properties, it should be noted that scanning surface harmonic microscopy [*Michel* et al. (1992)] has provided contrast in images of technological dopant patterns [*Bourgoin* et al. (1994)]. Here the resolution currently remains limited to the order of 100 nm.

In another technique derived from magnetic force microscopy, it has been demonstrated that ultimately high sensitivities for ESR [*Rugar* et al. (1992)] and NMR [*Rugar* et al. (1994)] can be achieved. Two-dimensional force maps enabled the reconstruction of spin density images with micrometer-scale spatial resolution [*Züger* and *Rugar* (1993, 1994)]. Three-dimensional imaging has also been demonstrated using this technique [*Züger* et al. (1996)]. The expected higher sensitivity of microfabricated force sensors should allow access to fewer than 100 individual spins [*Rugar* et al. (1994)]. Similar to the case of single photon detection, these are very impressive achievements. It remains, however, to be shown whether questions concerning the chemical or physical properties on the atomic or molecular scale can be addressed solely with these novel techniques.

2.4 Conclusion and Outlook

We have presented an extensive overview of how scanning probe microscopy techniques have contributed to our knowledge of chemical composition and reactivity. Uniquely, their resolution permits one to extract chemical and physical information or to follow chemical and physical processes down to the atomic and molecular scales. All the methods described are based on contrast mechanisms that are the secondary result of the chemical compositions. Most other analytical methods, in contrast, are based on the analysis of direct chemical signatures and offer straightforward interpretation. Mass spectrometry, for example, determines the exact concentrations of atomic nuclei. However, conventional chemical analysis techniques are often limited by a lack of spatial resolution. In contrast, SPM-based techniques are localized and noninvasive, and thereby allow chemical mapping with ultimate spatial sensitivity. As the SPM analytical experiments typically are performed with extremely small amounts of matter, signal-to-noise ratio and resolution represent much more challenging problems than in many of the conventional techniques in which larger sample volumes are averaged.

As a result of this ultimate resolution and sensitivity, we foresee further important developments in chemical imaging by scanning probe microscopy as more chemically specific transduction mechanisms are incorporated. The current trend is to explore new spectroscopic contrast mechanisms as the experimental sensitivity continues to improve. AFM, for example, has only recently been proved to achieve "true" atomic contrast (see references cited in the introduction to 2.4, above). Many of the new contrast mechanisms presented here demonstrate potential for generalized applicability. Progress in the spectroscopic characterization of individual molecules will allow experiments with single-molecular sensitivity. Single-atom and molecular spectroscopy have been performed within crystals [*Basché* et al. (1995), *Moerner* (1988, 1994), *Moerner* and *Basché* (1993)], on surfaces [*Betzig* and *Chichester* (1993), *Trautman* et al. (1994)] as well as in liquids [*Funatsu* et al. (1995)]. The combination of sensitive spectroscopic characterization with lateral confinement in microneedles [*Ishijima* et al. (1991)], optical traps [*Finer* et al. (1994)], and in SPM junctions allows chemical analysis at the ultimate limits of sensitivity and resolution.

Other offsprings of SPM are systems that combine micromechanical sensors with nanoscale processes. Although they have not been used for explorations in the context of imaging, a highly sensitive "nose" [*Gimzewski* et al. (1994)], a calorimeter [*Berger* et al. (1996)] and a thermal desorption spectrometer have been developed. On the other hand, molecular recognition and binding have been probed using force microscopes for such model systems as cell adhesion preoteoglycans [*Dammer* et al. (1995)], avidin–biotin [*Florin* et al. (1994)], complementary strands of DNA [*Lee* et al. (1994a)], and streptavidin–biotin [*Lee* et al. (1994b), *Moy* et al. (1994)]. It is interesting to speculate whether these biochemical recognition mechanisms used in biosensors [*Eggins* (1996)] can be combined with the above-mentioned, highly sensitive techniques. Ultimately, questions like the following will arise: Can molecular positioning [*Jung* et al. (1996)] or other technologies for nanoscale patterning (such as microcontact printing and advanced lithographies) help mankind to engineer and build functional molecular assemblies? Can molecular recognition mechanisms be borrowed from life's sensors and amplified to count the occupation of molecular receptors? Can molecular action be understood in its very detail, and will this enable us to integrate a more universal chemical sensing/mapping tool, going beyond the many examples presented to date and reviewed in this chapter?

In the context of the spectroscopic techniques presented in the first part of this chapter, similar questions arise, but concerning a smaller, atomic scale. The sensing limits and the reliability to build well-defined spectroscopic sensors will determine whether the methods demonstrated for elemental identification will be applicable to other material systems. Alternative approaches still have to be identified, especially for insulators.

Acknowledgements

We thank P. Guéret and H. Rohrer for enlightening and helpful discussions. During the preparation of this chapter, it was a great pleasure to conduct pertinent discussions with the many contributors to the field. In particular we would like to acknowledge those who provided figures and citations, often before final publication: R. Allensbach and U. Dürig; Ph. Avouris; G. Binnig; M. Bode, R. Wiesendanger and R. Berger; R. Berndt; A. Davies, J. Stroscio, D. Pierce and R.J. Celotta; E. Delamarche; D.M. Eigler and M. Crommie;

R. Hamers; D.D. Koleske, W.R. Barger, G.U. Lee and R.J. Colton; A. Marti, G. Hähner and N.D. Spencer; E. Meyer, R.M. Overney and R. Lüthi; J. Nogami; C. Padeste and L. Tiefenauer; M. Pfister, M.B. Johnson, S.F. Alvarado and H.W.M. Salemink; M. Schmid and P. Varga, E. Williams and O. Züger. It is a real pleasure to acknowledge IBM's publication department, especially C. Bolliger, U. Bitterli, D. Kropaci and L.-H. Pavka, for their professional support and assistance throughout the preparation of this chapter.

This work was partially supported by grants from the Swiss Federal Office for Education and Science through the ESPRIT basic research project PRONANO (8523) and by NSF under Award Nos. DMR-9624753 and DMR-9632527.

References

Allenspach R., Dürig U. (1996): private communication
Allenspach R., Bischof A., Dürig U. (1997): Surf. Sci. Lett. **381**, L573
Alvarado S.F. (1995): Phys. Rev. Lett. **75**, 513
Avouris Ph. (1991): In: *Highlights in Condensed Matter Physics and Future Prospects*, ed. by L. Esaki (Plenum, New York) p. 513
Avouris Ph., Lyo I.-W. (1994): Science **264**, 942
Barger W.R., Koleske D.D., Feldman K., Krüger D., Colton R.J. (1996): Polymer Preprints **37**, 606 (The Division of Polymer Chemistry, ACS, Washington, D.C.)
Basché Th., Kummer S., Bräuchle C. (1995): Nature **373**, 132
Batson P.E. (1993): Nature **366**, 727
Batson P.E. (1996): J. Electron. Microsc. **45**, 51
Becker R.S., Golovchenko J.A., Swartzentruber B.S. (1985a): Phys. Rev. Lett. **55**, 987
Becker R.S., Golovchenko J.A., Hamann D.R., Swartzentruber B.S. (1985b): Phys. Rev. Lett. **55**, 2032
Bedrossian P., Meade R.D., Mortensen K., Chen D.M., Golovchenko J.A., Vanderbilt D. (1989): Phys. Rev. Lett. **63**, 1257
Bell L.D., Kaiser W.J. (1988): Phys. Rev. Lett. **61**, 2368
Berger R., Gerber Ch., Gimzewski J.K., Meyer E., Güntherodt H.-J. (1996): Appl. Phys. Lett. **69**, 40
Berndt R., Gimzewski J.K. (1992): Phys. Rev. B **45**, 14095
Berndt R., Gimzewski J.K. (1993): Phys. Rev. B **48**, 4746
Berndt R., Gimzewski J.K., Johansson P. (1991): Phys. Rev. Lett. **67**, 3796
Berndt R., Gimzewski J.K., Schlittler R.R. (1992): Ultramicrosocpy **42–44**, 355
Berndt R., Gaisch R., Gimzewski J.K., Reihl B., Schlittler R.R., Schneider W.D., Tschudy M. (1993a): Science **262**, 1425
Berndt R., Gimzewski J.K., Johansson P. (1993b): Phys. Rev. Lett. **71**, 3493
Berndt R., Gaisch R., Schneider W.D., Gimzewski J.K., Reihl B., Schlittler R.R., Tschudy M. (1995): Phys. Rev. Lett. **74**, 102
Betzig E., Chichester R.J. (1993): Science **262**, 1422
Betzig E., Trautman J.K. (1992): Science **257**, 189
Betzig E., Lewis A., Harootunian A., Isaacson M., Kratschmer E. (1986): Biophys. J. **49**, 269
Betzig E., Isaacson M., Lewis A. (1987): Appl. Phys. Lett. **51**, 2088
Biebuyck H.A., Larsen N.B., Delamarche E., Michel B. (1997): IBM J. Res. Develop. **41**, 159

Biedermann A., Genser O., Hebenstreit W., Schmid M., Redinger J., Podloucky R., Varga P. (1996): Phys. Rev. Lett. **76**, 4179

Binnig G., Rohrer H., Gerber Ch., Weibel E. (1983): Phys. Rev. Lett. **50**, 120

Binnig G., Frank K.H., Fuchs H., Garcia N., Reihl B., Rohrer H., Salvan F., Williams A.R. (1985a): Phys. Rev. Lett. **55**, 991

Binnig G., Garcia N., Rohrer H. (1985b): Phys. Rev. B **32**, 1336

Bode M., Pascal R., Wiesendanger R. (1996a): Appl. Phys. A **62**, 571

Bode M., Pascal R., Dreyer M., Wiesendanger R. (1996b): Phys. Rev. B **54**, R8385

Bourgoin J.-P., Johnson M.B., Michel B. (1994): Appl. Phys. Lett. **65**, 2045

Browning N.D., Chisholm M.F., Pennycook S.J. (1993): Nature **366**, 143

Chambliss D.D., Chiang S. (1992): Surf. Sci. Lett. **264**, L187

Chander M., Goetsch D.A., Aldao C.M., Weaver J.H. (1995): Phys. Rev. Lett. **74**, 2014

Chavy C., Joachim C., Altibelli A. (1993): Chem. Phys. Lett. **214**, 569

Chen D.M., Golovchenko J.A., Bedrossian P., Mortensen K. (1988): Phys. Rev. Lett. **61**, 2867

Chiang S. (1992): "Molecular Imaging by Scanning Tunneling Microscopy", in *Scanning Tunneling Microscopy I*, ed. by H.-J. Güntherodt and R. Wiesendanger, 2nd edn., Springer Ser. Surf. Sci., Vol. 20 (Springer, Berlin, Heidelberg) pp. 181–205, and references therein

Coombs J.H., Gimzewski J.K. (1988): J. Microscopy **152**, 841

Crommie M.F., Lutz C.P., Eigler D.M. (1993a): Phys. Rev. B **48**, 2851

Crommie M.F., Lutz C.P., Eigler D.M. (1993b): Nature **363**, 524; Science **262**, 218

Dammer U., Popescu O., Wagner P., Anselmetti D., Güntherodt H.-J., Misevic G.N. (1995): Science **267**, 1173

Davies A., Stroscio J.A., Pierce D.T., Celotta R.J. (1996): Phys. Rev. Lett. **76**, 4175

Delamarche E., Michel B., Biebuyck H.A., Gerber Ch. (1996): Adv. Mater. **8**, 719

Dose V. (1985): Surf. Sci. Rep. **5**, 337

Doyen G., Drakova D. (1993): "The Scattering Theoretical Approach to Scanning Tunneling Microscopy and Scanning Tunneling Spectroscopy", in *Scanning Tunneling Microscopy III: Theory of STM and Related Scanning Probe Methods*, ed. by R. Wiesendanger and H.-J. Güntherodt, 2nd edn., Springer Ser. Surf. Sci., Vol. 29 (Springer, Berlin, Heidelberg) pp. 362–372

Drakova D., Doyen G. (1996): Surf. Sci. **352/354**, 698

Dürig U., Pohl D.W., Rohner F. (1986): J. Appl. Phys. **59**, 3318

Dürig U., Züger O., Pohl D.W. (1990): Phys. Rev. Lett. **65**, 349

Echenique P.M., Pendry J.B. (1990): Progr. Surf. Sci. **32**, 111

Eggins B.R. (1996): *Biosensors: an Introduction* (Wiley, Chichester & Teubner, Stuttgart)

Ehrlich G. (1992): Appl. Phys. A **55**, 403. In the field ion microprobe, it is possible to chemically analyze individual atoms by field-desorbing them. Such a procedure will, however, destroy the surface structure

Eigler D. (1994): "Quantum Corrals", in: *Nanostructures and Quantum Effects*, ed. by H. Sakaki and H. Nogewa, Springer Ser. Mater. Sci., Vol. 31 (Springer, Berlin, Heidelberg) pp. 311–314

Feenstra R.M. (1994): Surf. Sci. **299/300**, 965

Feenstra R.M., Stroscio J.A., Fein A.P. (1987): Surf. Sci. **181**, 295

Feldman K., Tervoort T., Smith P., Spencer N.D. (1997): Personal communication, and Langmuir (in press)

Finer J.T., Simmons R.M., Spudich J.A. (1994): Nature **368**, 113

Fischer R., Schuppler S., Fischer N., Fauster Th., Steinmann W. (1993a): Phys. Rev. Lett. **70**, 654

Fischer R., Fauster Th., Steinmann W. (1993b): Phys. Rev. B **48**, 15496

Florin E.-L., Moy V.T., Gaub H.E. (1994): Science **264**, 415

Frank E.R., Chen X.X., Hamers R.J. (1995): Surf. Sci. **334**, L709

Frisbie C.D., Rozsnyai L.F., Noy A., Wrighton M.S., Lieber C.M. (1994): Science **265**, 2071

Frommer J. (1992): Angew. Chem. Int. Ed. Engl. **104**, 1298

Fuchs H. (1993): J. Molec. Struc. **292**, 29

Funatsu T., Harada Y., Tokunaga M., Saito K., Yanagida T. (1995): Nature **374**, 555

Giessibl F.J. (1995): Science **267**, 68

For a review, see Gimzewski J.K. (1995): in *Photons and Local Probes*, ed. by O. Marti and R. Möller, NATO ASI Ser. E: Applied Sciences, Vol. 300 (Kluwer, Dordrecht) as well as the other contributions in that volume

Gimzewski J.K., Reihl B., Coombs J.H., Schlittler R.R. (1988): Z. Phys. B **72**, 497

Gimzewski J.K., Berndt R., Schlittler R.R. (1991): Ultramicrosc. **42/44**, 366

Gimzewski J.K., Gerber Ch., Meyer E., Schlittler R.R. (1994): Chem. Phys. Lett. **217**, 589

Goldstein J.I., Newbury D.E., Echlin P., Joy D.C., Fiori C., Lifshin E. (1981): *Scanning Electron Microscopy and X-Ray Microanalysis* (Plenum, New York)

Gould S., Marti O., Drake B., Hellemans L., Bracker C.E., Hansma P.K., Keder N.L., Eddy M.M., Stucky G.D. (1988): Nature **332**, 332

Gundlach K.H. (1966): Solid State Electron. **9**, 949

Güntherodt H.-J., Anselmetti D., Meyer E., Eds. (1995): *Forces in Scanning Probe Methods*, NATO ASI Ser. E: Applied Sciences, Vol. 286 (Kluwer, Dordrecht)

Hähner G., Marti A., Spencer N.D. (1997): Tribol. Lett. **3**, 359.

Hallmark V., Chiang S., Brown J.K., Woell Ch. (1991): Phys. Rev. Lett. **66**, 48

Hallmark V.M., Chiang S., Meinhardt K.P., Hafner K. (1993): Phys. Rev. Lett. **70**, 3740

Hamers R.J. (1989): Phys. Rev. B **40**, 1657

Hamers R.J. (1992): "STM on Semiconductors", in *Scanning Tunneling Microscopy I*, ed. by H.-J. Güntherodt and R. Wiesendanger, 2nd edn., Springer Ser. Surf. Sci., Vol. 20 (Springer, Berlin, Heidelberg) pp. 83–129

Hamers R.J., Demuth, J.E. (1988): Phys. Rev. Lett. **60**, 2527

Hamers R.J., Tromp R.M., Demuth J.E. (1986): Phys. Rev. Lett. **56**, 1972

Heinzelmann H., Pohl D.W. (1994): Appl. Phys. A **59**, 89

Himpsel F.J. (1990a): Phys. Scripta T **31**, 171

Himpsel F.J. (1990b): Surf. Sci. Rep. **12**, 1

Himpsel F.J., Ortega J.E. (1994): Phys. Rev. B **50**, 4992

Himpsel F.J., Mo Y.W., Jung T., Ortega J.E., Mankey G.J., Willis R.F. (1994): Superlattices and Microstructures **15**, 237

Himpsel F.J., Jung T.A., Schlittler R.R., Gimzewski J.K. (1996): Jpn. J. Appl. Phys., Pt. 1, **35**, 3695

Himpsel F.J., Ortega J.E., Mankey G.J., Willis R.F. (1998a): Adv. Physics (in press)

Himpsel F.J., Jung T., Seidler P.F. (1998b): IBM J. Res. Develop. **42**, Jan. 1, 1998

Hipps K.W., Mazur U. (1993): J. Phys. Chem. **97**, 7803

Hipps K.W., Lu X., Wang X.D., Mazur U. (1996): J. Phys. Chem. **100**, 11207

Holliday K., Wild U.P. (1993): In *Spectral Hole-Burning, Molecular Luminescence Spectroscopy*, ed. by St. G. Schulman, Chemical Analysis Ser., Vol. 77 (Wiley, New York) p. 149

Hölzl J., Schulte F.K. (1979): In *Work Function of Metals*, Springer Tracts Mod. Phys., Vol. 85 (Springer, Berlin, Heidelberg)

Howald L., Lüthi R., Meyer E., Günther, P., Güntherodt H.-J. (1994): Z. Phys. B **93**, 267

Ishijima A., Doi T., Sakurada K., Yanagida Y. (1991): Nature **352**, 301

Jesson D.E., Pennycook S.J., Baribeau L.-M. (1991): Phys. Rev. Lett. **66**, 750

Johansson P., Monreal R., Apell P. (1990): Phys. Rev. B **42**, 9210

Johnson M.B., Albrektsen O., Feenstra R.M., Salemink H.W.M. (1993): Appl. Phys. Lett. **63**, 2923

Johnson M.B., Koenraad P.M., van der Vleuten W.C., Salemink H.W.M., Wolter J.H. (1995): Phys. Rev. Lett. **75**, 1606

Jung T.A., Mo, Y.W., Himpsel, F.J. (1995a): Phys. Rev. Lett. **74**, 1641

Jung T.A., Schlittler R.R., Gimzewski J.K., Himpsel F.J. (1995b): Appl. Phys. A **61**, 467

Jung T.A., Schlittler R.R., Gimzewski J.K., Tang H., Joachim C. (1996): Science **271**, 181

Jung T.A., Schlittler R.R., Gimzewski J.K. (1997a): preprint

Jung T.A., Schlittler R.R., Gimzewski, J.K. (1997b): in preparation

Kalotas T.N., Lee A.R., Liesegang J., Alexopoulos A. (1996): Appl. Phys. Lett. **69**, 1710

Kazmerski L.L. (1991): J. Vac. Sci. Technol. B **9**, 1549

Koleske D.D., Barger W.R., Lee G.U., Colton R.J. (1997): Mat. Res. Soc. Symp. Proc. **464**, 377

Kumar A., Whitesides G.M. (1993): Appl. Phys. Lett. **63**, 2002

Kumar A., Biebuyck H.A., Whitesides G.M. (1994): Langmuir **10**, 1498

Lee G.U., Chrisey L.A., Colton R.J. (1994a): Science **266**, 771

Lee G.U., Kidwell D.A., Colton R.J. (1994b): Langmuir **10**, 354

Lu X., Hipps K.W., Wang X.D., Mazur U. (1996): J. Am. Chem. Soc. **118**, 7197

Lyo I.-W., Kaxiras E., Avouris Ph. (1989): Phys. Rev. Lett. **63**, 1261

Magonov S.N., Whangbo M.-H. (1996): *Surface Analysis with STM and AFM* (VCH, Weinheim)

Marti O., Colchero J., Mlynek J. (1990): Nanotechnology **1**, 141

Marti A., Hähner G., Spencer N.D. (1995): Langmuir **11**, 4632

Marti A., Hähner G., Spencer N.D. (1996a): reported at the Symposium Friction and Wear of the German Society for Material Science, Bad Nauheim, Germany (1996)

Marti A., Hähner G., Spencer N.D. (1996b): Poster presented at the 43rd Nat. Symp. American Vacuum Society, Philadelphia, PA (October 1996)

Meyer G., Amer N.M. (1990): Appl. Phys. Lett. **57**, 2089

Meyer E., Heinzelmann H. (1992): "Scanning Force Microscopy (SFM)", in *Scanning Tunneling Microscopy II: Further Applications and Related Scanning Techniques*, ed. by R. Wiesendanger and H.-J. Güntherodt, 2nd edn., Springer Ser. Surf. Sci., Vol. 28 (Springer, Berlin, Heidelberg) pp. 99–149

Meyer E., Howald L., Overney R.M., Heinzelmann H., Frommer J., Güntherodt H.-J., Wagner T., Schier H., Roth S. (1991): Nature **349**, 398

Meyer E., Overney R., Lüthi R., Brodbeck D., Howald L., Frommer J., Güntherodt H.-J., Wolter O., Fujihira M., Takano H., Gotoh Y. (1992): Thin Solid Films **220**, 132

Michel B., Mizutani W., Schierle R., Jarosch A., Knop W., Benedikter H., Bächtold W., Rohrer H. (1992): Rev. Sci. Instrum. **63**, 4080

Mo Y.W. (1992): Phys. Rev. Lett. **69**, 3643

Mo Y.W. (1993a): Science **261**, 886

Mo Y.W. (1993b): Phys. Rev. Lett. **71**, 2923

Mo Y.W., Himpsel F.J. (1994): Phys. Rev. B **50**, 7868

For a review, see (a) Moerner W.E. (Ed.) (1988): *Persistent Spectral Hole-Burning: Science and Applications*, Topics Curr. Phys., Vol. 44 (Springer, Berlin, Heidelberg); (b) Moerner W.E. (1994): Science **265**, 361; (c) Moerner W.E., Basché Th. (1993): Angew. Chem. **105**, 537; Angew. Chem. Int. Ed. Eng. **32**, 457

Moers M.H.P, Gaub H.E., van Hulst N.F. (1994): Langmuir **10**, 2774

Morozov V.N., Seeman N.C., Kallenbach N.R. (1993): Scanning Microsc. **7**, 757

Morrison S.R. (1990): *The Chemical Physics of Surfaces*, 2nd edn. (Plenum, New York)

Moy V.T., Florin E.-L., Gaub H.E. (1994): Science **266**, 257

Moyer P.J., Walzer K., Hietschold M. (1995): Appl. Phys. Lett. **67**, 2129

Müller E.W., Tsong T.T. (1969): *Field Ion Microscopy Principles and Applications* (Elsevier, New York)

Muller D.A., Tzou Y., Raj R., Silcox J. (1993): Nature **366**, 725

Newbury Dale E. (1990): Nanotechnology **1**, 103

Nogami J. (1994): Surf. Rev. Lett. **1**, 395

Onishi H., Iwasawa Y. (1996): Phys. Rev. Lett. 76, 791

O'Shea J.J., Brazel E.G., Rubin M.E., Bhargava S., Chin M.A., Narayanamurti V. (1997): Phys. Rev. B **56**, 2026

Overney R.M., Meyer E., Frommer J., Brodbeck D., Luethi R., Howald L., Günthe-rodt H.-J., Fujihira M., Takano H., Gotoh Y. (1992a): Nature **359**, 133

Overney R.M., Howald L., Frommer J., Meyer E., Brodbeck D., Güntherodt H.-J. (1992b): Ultramicroscopy **42/44**, 983

Overney R.M., Meyer E., Frommer J., Güntherodt H.-J., Fujihira M., Takano H., Gotoh Y. (1994): Langmuir **10**, 1281

Paesler M.A., Moyer P.J. (1996): *Near-Field Optics. Theory, Instrumentation, and Applications* (Wiley, New York)

Pechman R.J., Wang X.-S., Weaver J.H. (1995): Phys. Rev. B **52**, 11412

Pennycook S.J., Boatner L.A. (1988): Nature **336**, 565

Pennycook S.J., Jesson D.E. (1990): Phys. Rev. Lett. **64**, 938

Persson B.N.J., Baratoff A. (1992): Phys. Rev. Lett. **68**, 3224

Pfister M., Johnson M.B., Alvarado S.F., Salemink H.W.M., Marti U., Martin D., Morier–Genoud F., Reinhart F.K. (1995): Appl. Phys. Lett. **67**, 1459

Pitarke J.M., Flores F., Echenique P.M. (1990): Surf. Sci. **234**, 1

Pohl D.W. (1991): "Scanning Near-field Optical Microscopy (SNOM)", in *Advances in Optical and Electron Microscopy*, **12**, 243 (Academic, London)

Pohl D.W. (1992): "Nano-Optics and Scanning Near-Field Optical Microscopy", in *Scanning Tunneling Microscopy II: Further Applications and Related Scanning Techniques*, ed. by R. Wiesendanger and H.-J. Güntherodt, 2nd edn., Springer Ser. Surf. Sci., Vol. 28 (Springer, Berlin, Heidelberg) pp. 233–271

Pohl D.W., Denk W., Lanz M. (1984): Appl. Phys. Lett. **44**, 651

Pohl D.W., Denk W., Dürig U. (1985): Proc. SPIE **565**, 56

Quate C.F. (1994): Surf. Sci. **299/300**, 980

Reed S.J.B. (1975): *Electron Microprobe Analysis* (Cambridge Univ. Press, London)

Reimer L. (1985): *Scanning Electron Microscopy. Physics of Image Formation and Microanalysis*, 2nd edn. Springer Ser. Opt. Sci., Vol. 45 (Springer, Berlin, Heidel-berg)

Rettenberger A., Baur C., Läuger K., Hoffmann D., Grand J.Y., Möller R. (1995): Appl. Phys. Lett. **67**, 1217

Ritz G., Schmid M., Biedermann A., Varga P. (1996): Phys. Rev. B **53**, 16019

Roberts G. (1990): *Langmuir Blodgett Films* (Plenum, New York)

Rohrer H. (1994): Surf. Sci. **299/300**, 956

Ruan L., Besenbacher F., Stensgaard I., Laegsgaard E. (1993): Phys. Rev. Lett. **70**, 4079

Rugar D., Yannoni C.S., Sidles J.A. (1992): Nature **360**, 563

Rugar D., Züger O., Hoen S., Yannoni C.S., Vieth H.-M., Kendrick R.D. (1994): Science **264**, 1560

Salemink H.W.M., Albrektsen O. (1993): Phys. Rev. B **47**, 16044

Sautet P., Joachim C. (1991): Chem. Phys. Lett. **185**, 23

Sautet P., Joachim C. (1992): Surf. Sci. **271**, 187

Schmid M.: Personal communication

Schmid M., Stalder H., Varga P. (1993): Phys. Rev. Lett. **70**, 1441

Schönenberger C., Jorritsma J., Sondag–Huethorst J.A.M., Fokkink L.G.J. (1995): J. Phys. Chem. **99**, 3259

Sirringhaus H., Lee E.Y., von Känel H. (1994): Phys. Rev. Lett. **73**, 577

Sirringhaus H., Lee E.Y., von Känel H. (1995): Phys. Rev. Lett. **74**, 3999

Smith N.V. (1988): Rep. Progr. Phys. **51**, 1227

Sprik M., Delamarche E., Michel B., Röthlisberger U., Klein M.L., Wolf H., Ringsdorf H. (1994): Langmuir **10**, 4116

Steinmann W., Fauster Th. (1995): "Two-Photon Photoelectron Spectroscopy of Electronic States at Metal Surfaces," Chap. 5, in: *Laser Spectroscopy and Photochemistry on Metal Surfaces*, ed. by H.-L. Dai and W. Ho, Advanced Series in Physical Chemistry, Vol. 5 (World Scientific, Singapore, 1995) p. 184

Stroscio J.A., Pierce D.T., Davies A., Celotta R.J., Weinert M. (1995): Phys. Rev. Lett. **75**, 2960

Sugawara Y., Ohta M., Ueyama H., Morita S. (1995): Science **270**, 1646

Takami T., Delamarche E., Michel B., Gerber Ch., Wolf H., Ringsdorf H. (1995): Langmuir **11**, 3876

Tao N.J. (1996): Phys. Rev. Lett. **76**, 4066

Tersoff J., Hamann D.R. (1985): Phys. Rev. B **31**, 805

Trautman J.K., Macklin J.J., Betzig E. (1994): Nature **369**, 40

Tsong T.T. (1990): *Atom-Probe Field Ion Microscopy* (Cambridge Univ. Press, Cambridge)

Uehara Y., Kimura Y., Ushioda S., Takeuchi K. (1992): Jpn. J. Appl. Phys. **31**, 2465

Ulman A. (1991): *An Introduction to Ultrathin Organic Films From Langmuir Blodgett to Self-Assembly* (Academic, San Diego, CA) and references therein

van der Werf K.O., Putman C.A.J., de Grooth B.G., Greve J. (1994): Appl. Phys. Lett. **65**, 1195

Vázquez de Parga A.L., Hernán O.S., Miranda R., Levy Yeyati A., Mingo N., Martín-Rodero A., Flores F. (1998): Phys. Rev. Lett. **80**, 357

Viernow J., Lin J.L., Petrovykh D.Y., Men F.K., Seo D.J., Henzler M., Himpsel F.J. (1998): Phys. Rev. B, submitted

Wang Y., Chen X., Hamers R.J. (1994): Phys. Rev. B **50**, 4534

Wiesendanger R., Bode M., Pascal R., Allers W., Schwarz U.D. (1996): J. Vac. Sci. Technol. A **14**, 1161

Wilbur J.L., Kumar A., Kim E., Whitesides G.M. (1994): Adv. Mater. **6**, 600

Wilbur J.L., Biebuyck H.A., MacDonald J.C., Whitesides G.M. (1995): Langmuir **11**, 825

Williams C.C., Wickramasinghe H.K. (1990): Nature **344**, 317

Wintterlin J., Behm R.J. (1992): "Adsorbate Covered Metal Surfaces and Reactions on Metal Surfaces", in *Scanning Tunneling Microscopy I*, ed. by H.-J. Güntherodt and R. Wiesendanger, 2nd ed., Springer Ser. Surf. Sci., Vol. 20 (Springer, Berlin, Heidelberg) pp. 39–82

Wolf H., Ringsdorf H., Delamarche E., Takami T., Kang H., Michel B., Gerber Ch., Jaschke M., Butt H.-J., Bamberg E. (1995): J. Phys. Chem. **99**, 7102

Wolkow R., Avouris Ph. (1988): Phys. Rev. Lett. **60**, 1049

Woodruff D.P. (1994): J. Phys.: Condens. Matter **6**, 6094

Wouda P.T., Nieuwenhuys B.E., Schmid M., Varga P. (1996): Surf. Sci. **359**, 17

Zambelli T., Trost J., Wintterlin J., Ertl G. (1996): Phys. Rev. Lett. **76**, 795

Zenhausern F., Martin Y., Wickramasinghe H.K. (1995): Science **269**, 1083

Zheng J.F., Liu X., Newman N., Weber E.R., Ogletree D.F., Salmeron M.B. (1994a): Phys. Rev. Lett. **72**, 1490

Zheng J.F., Salmeron M.B., Weber E.R. (1994b): Appl. Phys. Lett. **64**, 1836

Züger O., Rugar D (1993): Appl. Phys. Lett. **63**, 2496

Züger O., Rugar D. (1994): J. Appl. Phys. **75**, 6211

Züger O., Hoen S.T., Yannoni C.S., Rugar D. (1996): J. Appl. Phys. **79**, 1881

3. Thermovoltages
in Scanning Tunneling Microscopy

R. Möller

With 15 Figures

A temperature gradient in a conducting material leads to a gradient of the electric field. The ratio of the two gradients is called thermopower or Seebeck coefficient. It results from the balance of transport phenomena and it is a characteristic of the material. If an electric circuit is formed by two different materials and a temperature difference between the two connections is present, a thermoelectric current will be observed for a closed loop, or a thermovoltage at open loop.

The thermovoltage, which will be discussed in the following, occurs when the two electrodes on both sides of the vacuum tunneling barrier of a Scanning Tunneling Microscope (STM) are at different temperatures. The situation is sketched in Fig. 3.1. If, e.g., the tip is heated this yields a temperature gra-

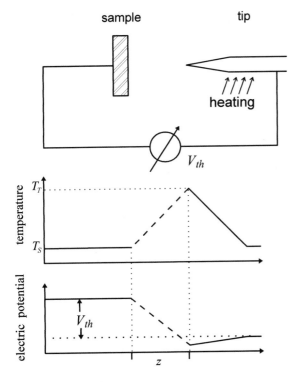

Fig. 3.1. Schematic representation of an experimental arrangement to investigate the thermovoltage across a vacuum tunneling barrier

dient within the tip and its support but the sample will stay at the ambient temperature since the thermal coupling across a vacuum barrier is negligible compared to the thermal conductance of the sample. To use the terminology of conventional thermovoltage, one may define a "thermopower" of the vacuum tunneling barrier which is the ratio between the thermovoltage and the temperature difference between tip and sample. However, one should not stress this analogy too much because this thermopower is not a characteristic property of the vacuum but it depends on the electronic states of the tip and sample, as will be discussed later. It can be as large as $80\,\mu V/K$ for metallic tunneling tip and sample, hence about one order of magnitude larger than the conventional thermopower for typical metals. It should be noted that the thermovoltage which is measured externally is the sum of the conventional thermovoltage along the heated tip and the thermovoltage of the vacuum barrier. Of course one may also heat or cool the sample instead of the tip.

The thermovoltage across the vacuum barrier results from the balance between forward and backward tunneling of electrons. In absence of a temperature difference, the two processes cancel exactly. However, if the Fermi distributions of both electrodes are different, this will lead to a net current for zero bias. If the external current is zero, this will induce an extra electric charge which is associated with a bias voltage. This voltage yields a "normal" tunneling current which is opposed to the thermally driven current. At equilibrium the two cancel exactly. Since tunneling may only occur at energies close to the Fermi-level in absence of an external bias voltage, the thermopower will depend on the local density of electronic states of the tip and the sample around the Fermi-level. Minor changes of the electronic structure may significantly modify the balance which is responsible for the thermovoltage across the vacuum barrier. Therefore, the latter may provide information about details of the local density of electronic states, which may not be easily obtained by other measurements, e.g., tunneling spectroscopy. This may be applied, e.g., for the investigation of electronic surface states or to distinguish different chemical elements on a heterogeneous surface.

The investigation of the thermoelectric effects in STM has been pioneered by *Williams* and *Wickramasinghe* [3.1] who investigated at ambient conditions the thermovoltage on a MoS_2 sample. They found the atomic periodicity as well in the topography as in the thermovoltage. However, while the maxima in the topography corresponded to the Mo-sublattice the maxima in the thermovoltage coincided with the S-sublattice.

3.1 Theory

A theoretical description for the thermovoltage across a tunneling barrier was given by Kohler more than fifty years ago [3.2]. He considered a metal-oxide-metal junction which is formed if a thermocouple is built using oxidized wires. In context to STM the thermopower of a vacuum tunneling barrier has been treated by *Leavens* and *Aers* [3.3]. In the following, the approach by *Støvneng* and *Lipavský* [3.4] will be outlined, which is less complex and yields useful approximations.

The description is based on the *Tersoff-Hamann-approximation* [3.5]:

$$I = a \int \rho_{\mathrm{T}}(E + \mathrm{eV}/2)\rho_{\mathrm{S}}(\boldsymbol{r}_{\mathrm{T}}, E - \mathrm{eV}/2)$$

$$\times \left\{ f\left[\frac{(E - \mathrm{eV}/2)}{kT_{\mathrm{S}}}\right] - f\left[\frac{(E + \mathrm{eV}/2)}{kT_{\mathrm{T}}}\right] \right\} \mathrm{d}E \ , \tag{3.1}$$

where a is a constant, $\rho_{\mathrm{T}}(E)$ is the density of electronic states of the tunneling tip, $\rho_{\mathrm{S}}(\boldsymbol{r}_{\mathrm{T}}, E)$ is the density of electronic states of the sample at the position of the tip $\boldsymbol{r}_{\mathrm{T}}$, f is the Fermi-Dirac function, k is Boltzmann's constant, T_{S} and T_{T} are the temperatures of the sample and tip, respectively. In contrast to the standard form, the zero point for the energy has been chosen in the middle between the Fermi level of tunneling tip and sample. The product $\rho_{\mathrm{T}}\rho_{\mathrm{S}}$ results from the summation over the tunneling matrix elements [3.6]

$$M = \frac{\hbar}{2m} \int_{S} \left(\psi_{\mu}^{*} \nabla \psi_{\nu} - \psi_{\nu} \nabla \psi_{\mu}^{*} \right) \mathrm{d}S \tag{3.2}$$

for all possible combinations of electronic wave functions ψ_{μ} for the tunneling tip and ψ_{ν} for the sample. $\hbar$ is Planck's constant, m is the electron mass, S is an arbitrary surface anywhere between the tip and sample which covers the region of significant overlap of the wave functions (which is traversed by all "tunneling electrons"). Hence the matrix elements contain the decay of the wave functions within the tunneling barrier.

As *Støvneng* and *Lipavský* proposed, the discussion of the thermopower may be facilitated if the surface of the tip is used for S. For this particular choice only the exponential decay of the wave function of the sample has to be considered. If the square of $k_{\parallel}$, the component of the wave vector parallel to the surface, is small compared to $(2m/\hbar^2)(\phi - E)$, the density of states of the sample taken at the tip surface is of the form

$$\rho(\boldsymbol{r}_{\mathrm{T}}, E) = \rho(x, y, E) \exp\left[(-2/\hbar)\sqrt{2m(\phi - E)} \, z\right] \ , \tag{3.3}$$

where x, y and z are the coordinates of the tip and ϕ is the work function of the sample (at x, y).

Assuming that the involved bias voltage is small compared to kT, the expression for the tunneling current (3.1) can be linearly expanded to

$$I = I_{\mathrm{th}} + \sigma V \ , \tag{3.4}$$

where I_{th} is the thermally driven tunneling current if there is no bias voltage. It is given by:

$$I_{\mathrm{th}} = a \int \rho_{\mathrm{T}}(E)\rho_{\mathrm{S}}(\boldsymbol{r}_{\mathrm{T}}, E)[f(E/kT_{\mathrm{S}}) - f(E/kT_{\mathrm{T}})] \, \mathrm{d}E \ . \tag{3.5}$$

σ is the tunneling conductivity given by:

$$\sigma = -\frac{1}{2}ea \int \rho_{\mathrm{T}}(E)\rho_{\mathrm{S}}(\boldsymbol{r}_{\mathrm{T}}, E) \left[f'\left(\frac{E}{kT_{\mathrm{S}}}\right) + f'\left(\frac{E}{kT_{\mathrm{T}}}\right) \right] \mathrm{d}E \;, \tag{3.6}$$

where the prime denotes the derivative with respect to energy.

Experimentally the thermovoltage is observed at open loop conditions, hence at $I = 0$. This yields a thermovoltage at the tip

$$V_{\mathrm{th}} = -I_{\mathrm{th}}/\sigma \;. \tag{3.7}$$

The dependence of the thermovoltage can be understood qualitatively by looking at the difference of the Fermi functions which determines I_{th}, and the sum of their derivatives which determines σ. Both functions are plotted in Fig. 3.2 for $T_{\mathrm{S}} = 310\,\mathrm{K}$ and $T_{\mathrm{T}} = 300\,\mathrm{K}$.

If $T_{\mathrm{S}} > T_{\mathrm{T}}$ and if the product $\rho_{\mathrm{T}}(E)\rho_{\mathrm{S}}(\boldsymbol{r}_{\mathrm{T}}, E)$ increases monotonically (decreases) with increasing energy, the thermovoltage will be negative (positive), since σ is always positive. For example, if the energy dependence of $\rho_{\mathrm{T}}(E)$ and $\rho_{\mathrm{S}}(\boldsymbol{r}_{\mathrm{T}}, E)$ is caused only by the decay of the wave functions, which leads to a growing overlap with increasing energy, this yields a negative thermovoltage, which may be considered as the "normal" case.

The integrals for I_{th} and σ may be solved analytically if one assumes that $\rho_{\mathrm{T}}(E)\,\rho_{\mathrm{S}}(\boldsymbol{r}_{\mathrm{T}}, E)$ depends only linearly on the energy in the range which significantly contributes to the integral. This is about $\pm 8kT$ as can be seen in Fig. 3.2. Hence one may write:

$$\rho_{\mathrm{T}}(E)\rho_{\mathrm{S}}(\boldsymbol{r}_{\mathrm{T}}, E) = \rho_{\mathrm{T}}(0)\rho_{\mathrm{S}}(\boldsymbol{r}_{\mathrm{T}}, 0) + [\rho_{\mathrm{T}}(0)\rho_{\mathrm{S}}(\boldsymbol{r}_{\mathrm{T}}, 0)]'\, E \;. \tag{3.8}$$

The evaluation of the integrals yields

$$I_{\mathrm{th}} = a(\pi^2 k^2/6)\left(T_{\mathrm{S}}^2 - T_{\mathrm{T}}^2\right) \left[\rho_{\mathrm{T}}(0)\rho_{\mathrm{S}}(\boldsymbol{r}_{\mathrm{T}}, 0)\right]' \tag{3.9}$$

and

$$\sigma = ea\rho_{\mathrm{T}}(0)\rho_{\mathrm{S}}(\boldsymbol{r}_{\mathrm{T}}, 0) \;. \tag{3.10}$$

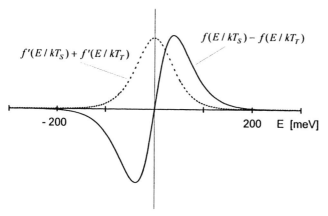

Fig. 3.2. Difference of the Fermi distribution for two different temperatures (*full line*), sum of the derivatives of the Fermi-distributions with respect to the energy (*dashed line*)

Hence one obtains for the thermovoltage:

$$V_{th} = \frac{\pi^2 k^2}{6e} \left(T_T^2 - T_S^2\right) \frac{[\rho_T(0)\rho_S(\mathbf{r}_T, 0)]'}{\rho_T(0)\rho_S(\mathbf{r}_T, 0)}$$

$$= \frac{\pi^2 k^2}{6e} \left(T_T^2 - T_S^2\right) [\ln \sigma(0)]' \ . \tag{3.11}$$

Expanding $(T_T^2 - T_S^2)$ into $(T_T - T_S)(T_T + T_S)$ shows that the thermovoltage depends linearly on the temperature difference (for constant mean temperature), as one would expect. The expression for the thermovoltage can be further modified to separate the different contributions using the above approximation for $\rho(\mathbf{r}_T, E)$:

$$V_{th} = \frac{\pi^2 k^2}{6e} \left(T_T^2 - T_S^2\right) \left\{ [\ln \rho_T(0)]' + [\ln \rho_S(x, y, 0)]' + \frac{z}{\hbar}\sqrt{\frac{2m}{\phi}} \right\} \ . \tag{3.12}$$

Hence the properties of the tunneling barrier is contained in a sum of three terms: The first and the second terms are given by logarithmic derivatives of the density of electronic states for tip and sample, respectively. While the first may be assumed to be constant the second term contains the lateral variation if the tip is scanning over the surface. The third contribution results from the energy dependence of the exponential decay of the wave functions. It depends on the tip-sample-distance and the work function which also might vary for different positions over the sample. It yields $3.76 z \sqrt{1/\phi}$ [μV/K] if z is given in [Å] and if the mean temperature is $300\,$K.

3.2 Experimental Setup

The thermovoltage across a tunneling barrier can in principle be investigated by many different experimental setups. However, it is very appealing to use the experimental configuration of an STM mainly for two reasons: First of all it allows a well defined variation of the tunneling barrier since the tip can be moved normal and lateral to the surface. A second advantage is the negligible thermal coupling between tip and sample if the experiments are performed at UHV conditions [3.7]. For the purpose of the experiment a conventional setup for STM has to be modified in two details. First, to generate the temperature difference between tip and sample, heating or cooling of either the tip or the sample is required. Second, the electronic control of the STM has to be extended to enable the measurement of the thermovoltage.

In the first experiments *Williams* and *Wickramasinghe* [3.1] used an electric heater attached to the sample, thereby achieving a temperature of $10\,$K above ambient. In our experiment the tip is heated by means of a small laser focused on the shaft of the tunneling tip. Since the thermal coupling to the rest of the setup of the latter is much less than for the sample, a power of $40\,$mW is sufficient to create a temperature difference of $5\,$K. This method is

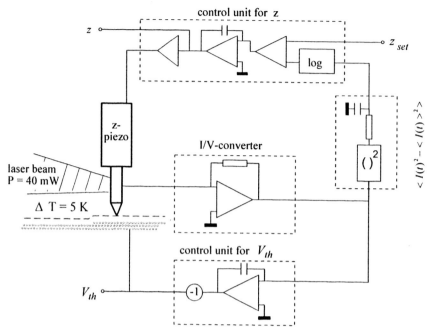

Fig. 3.3. Experimental setup for the investigation of the thermovoltage in STM

convenient for several reasons, e.g., it may be easily modulated, there is no electrical cross talk, etc. A scheme of the setup is shown in Fig. 3.3.

Several schemes have been developed to measure the difference of the electrical potential between tip and sample. These techniques are usually named potentiometry. Two methods to control the gap width as well as the voltage may be distinguished. For the first method, the tip (or the sample) is alternatingly connected to measurement of the tunneling current or the gap voltage. Hence, the two quantities are measured sequentially. The other group of techniques uses a simultaneous measurement of the tunneling conductivity and the bias voltage. This can be done by separating the ac- and the dc-component of the tunneling current. This has been first proposed by *Muralt* and *Pohl* [3.8]. Further refinement was introduced by *Kirthley* et al. [5.9] and by *Pelz* et al. [5.10]. In our group, a technique has been developed for which the ac-component of the tunneling current is not due to an external bias voltage but to the thermal noise of the tunneling conductivity. The virtue of this technique is the excellent resolution for the bias voltage between tip and sample.

The method is readily understood by looking at the tunneling current I as function of time for a given tunneling conductance and bias voltage:

$$I(t) = \sigma(V_{\text{th}} + V_{\text{ext}}) + a(t)\sqrt{4kTB\sigma} ,\qquad(3.13)$$

where V_{ext} stands for an external bias voltage, B for the bandwidth which is detected, $a(t)$ is a function with $\langle a(t) \rangle = 0$ and $\langle a(t)^2 \rangle = 1$ describing the statistical fluctuations.

The first term yields the normal tunneling current, while the second corresponds to the thermal fluctuations. The mean and the mean square value of I are:

$$\langle I(t) \rangle = \sigma(V_{\text{th}} + V_{\text{ext}}) \tag{3.14}$$

and

$$\langle I(t) - \langle I(t) \rangle^2 \rangle = 4kTB\sigma \propto \exp(-2\kappa z) \ . \tag{3.15}$$

The second expression is proportional to the tunneling conductance which depends exponentially on the distance z between tip and sample. Hence, it may be used for the feedback loop for the gap width instead of the tunneling current in normal STM operation. A constant value for $\langle I(t)^2 - \langle I(t) \rangle^2 \rangle$ corresponds to a fixed value of z assuming that κ remains constant.

The first term is used to evaluate V_{th} by adjusting V_{ext} such that $\langle I(t) \rangle$ vanishes. Hence, the bias which is applied externally becomes equal to the negative value of the thermovoltage (as long as $\sigma > 0$ which is guaranteed by the other feedback).

The scheme for the electronic circuitry is also shown in Fig. 3.3. The tunneling current is measured by means of a low noise I-V-converter connected to the tip (or the sample), providing a proportional output voltage. For the control of the gap width the variance of this value is electronically formed. This signal is fed into a conventional control unit for z which is normally used for "constant current" operation of the STM. Hence, the logarithm is evaluated which is proportional to the gap width. The difference to a preset value yields the error signal. It is fed into an integrator which drives the z-piezo via a high voltage amplifier. The output of the integrator provides a voltage which is proportional to the z-coordinate of the tip.

The lower part of the figure shows the feedback loop for the bias voltage. Since $\langle I(t) \rangle$ is proportional to the bias voltage it provides directly the error signal for the control unit which shall adjust the bias to zero. Since the integrator used for this purpose provides the necessary averaging, the output of the I/V-converter may be directly connected to its input. The output of the integrator is inverted (the gain of the complete loop must be negative) and connected to the sample (or the tip). At steady state condition it yields V_{ext}, hence the negative value of the thermovoltage.

3.3 Results

As can be seen from the above relation an almost linear current-to-voltage characteristic for small voltages is required for correct functioning of the technique used to measure the thermovoltage. To provide clean metallic surfaces, the samples and the tips have to be prepared in situ. All the experimental observations which will be presented were obtained under ultra high vacuum.

The tips were prepared by electrochemical etching of a polycrystalline tungsten wire of 0.5 mm in diameter. They were cleaned in situ by head on sputtering with Ar^+ ions of 2 keV at a current density of $150\,\mu A/cm^2$ for several hours. Shortly before the measurement an additional preparational step was performed by field desorption. The linearity of the current to voltage characteristic was verified on clean metal surfaces. Furthermore, it was checked that dI/dz corresponds to the appropriate work function of several eV.

The first observation of the thermopower of the vacuum barrier between metallic surfaces was made for polycrystalline films of noble metals. Many distinct features are found in the thermovoltage which exhibit a complicated correlation to the topography. Perhaps the most significant details were areas of constant low (high) thermovoltage found on flat terraces of some grains. From other investigations it was known that these facets are (111)-oriented. This suggests that the crystallographic orientation plays an important role and that the (111)-face which has the densest packing of atoms, represents one extreme value. Unfortunately, the structures on the polycrystalline material are too complex to provide a further understanding of the observation.

To obtain more insight into the origin of the lateral variation of the thermovoltage surfaces with well defined orientation were investigated. Films of Ag(111), Au(111) and Cu(111) were prepared by epitaxial growth on a mica substrate. The latter was heated to 550 K twelve hours before and during the evaporation. The metals were evaporated at a rate of 3 to 10 Å/s as determined by a quartz microbalance. A total thickness of 2000 Å was chosen, providing films which exhibit large flat terraces with (111) orientation. The parameters for the preparation agree with those reported by other groups [3.11].

Figure 3.4 shows the result for a Ag(111) surface. The topography displayed in the part a shows large terraces which are atomically flat. Several monoatomic step lines are clearly visible. The observed thermovoltage shown in the part b varies between −350 and −150 μV. It is essentially constant on the flat parts and every step yields a positive signal of about $\Delta V_{th} = 20\,\mu V$. To illustrate this effect in more detail, a line scan across two steps is displayed in Fig. 3.5. The upper trace shows the monoatomic steps in the topography. The transition from the lower to the upper terrace occurs laterally over about 0.5 nm (depending on the tip shape). The lower trace shows the corresponding thermovoltage. On the terrace the thermovoltage amounts to −300 μV whereas at the steps it shows a maximum. A close inspection reveals that it corresponds to the transition from the upper terrace to the lower one in the topography as indicated by the dashed lines. If the same area is investigated with the opposite scan direction for x an identical image is obtained for the topography as well as for the thermovoltage.

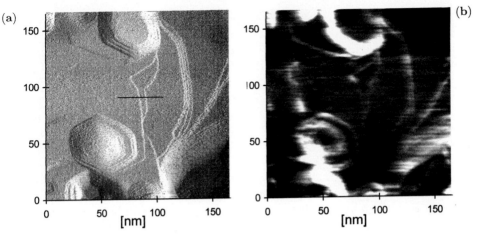

Fig. 3.4a, b. Topography and thermovoltage for Ag(111) on mica. The scan area is $165 \times 165 \, \text{nm}^2$, the tunneling conductance is held constant at about $10^{-8} \, \text{A/V}$. The *black line* indicates the position of the cross section displayed in Fig. 3.5. (**a**) topography (**b**) simultaneously measured thermovoltage ranging from -350 to $-150 \, \mu\text{V}$

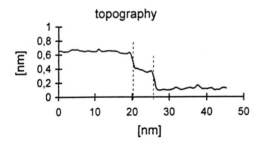

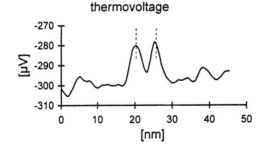

Fig. 3.5. Line scan across two monoatomic steps of Fig. 3.4. The *upper trace* shows the topography, the *lower trace* the corresponding thermovoltage

This particular behavior of the thermovoltage at step edges has been observed for different samples and tips. There is no indication that this signal depends on the orientation of steps. Since the signals occur at marked variations in the topography, one may suspect crosstalk between the feedback loop for the thermovoltage and the topography. However, this can be excluded for

(a)

(b)

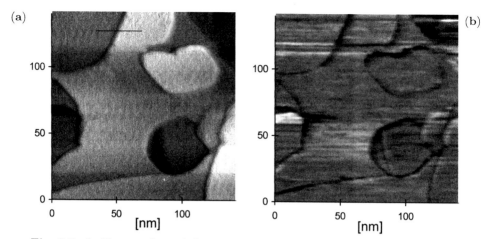

Fig. 3.6a, b. Topography and thermovoltage for Cu(111) on mica. The scan area is 140×140 nm, the tunneling conductance is held constant at about 10^{-8} A/V. The *black line* indicates the position of the cross section shown in Fig. 3.7. (**a**) topography varying by about 1 nm in total (**b**) simultaneously measured thermovoltage ranging from -200 to $-120\,\mu$V

several reasons. For example, if the laser which heats the tip is turned off, the signals in the thermovoltage vanish.

To learn more about the thermovoltage at step edges, the thermovoltage has been studied for the Cu(111) surface. The results are shown in Fig. 3.6. The parameters resemble those for the previous figure. The topography displayed in the part a shows large terraces which are atomically flat separated by a few monoatomic steps. The total corrugation is about 1 nm. The thermovoltage shown in the part a varies between -120 and $-200\,\mu$V. The variation of the sample height and thermovoltage at a step is again shown in a line scan in Fig. 3.7. Compared to Ag(111) the thermovoltage is remarkably different. The value on the flat terraces is about $-130\,\mu$V; the variation at the step is about $-30\,\mu$V. The latter has about the same magnitude as on silver but the opposite sign.

To analyse the variation of the thermovoltage at step edges in more detail, a small area of a Ag(111) surface has been investigated at higher resolution. The topography displayed in Fig. 3.8a shows one monoatomic step between two flat terraces. The latter exhibit the atomic corrugation characteristic for the Ag(111) surface. The part b of the figure shows the corresponding thermovoltage; the gray scale ranges between -320 and $-260\,\mu$V. In analogy to the data of Figs. 3.4, 5 the thermovoltage raises at the step edge by about $30\,\mu$V. The step edge appears very fuzzy because of the high mobility of the silver atoms at the step edge. However, there are a few lines where the step edge appears sharp and well defined. For these lines the lateral extension of the maximum in the thermovoltage is only about 0.3 nm. Hence, the large

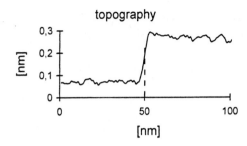

topography

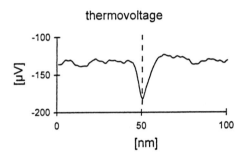

thermovoltage

Fig. 3.7. Line scan across a mono-atomic steps of Fig. 3.6. The *upper trace* shows the topography, the *lower trace* the corresponding thermovoltage

(a)

(b)

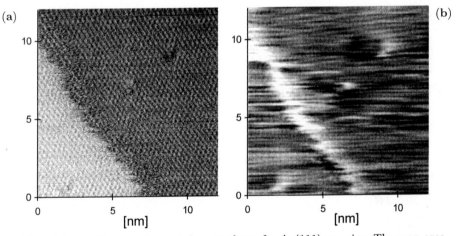

Fig. 3.8a, b. Topography and thermovoltage for Ag(111) on mica. The scan area is 12×12 nm, the tunneling conductance is held constant at about 10^{-8} A/V. The atomic structure of the Ag(111) surface can be recognized. (a) topography (b) simultaneously measured thermovoltage ranging from -350 to $-250\,\mu$V

width which has been observed on the larger scan is due to averaging over the diffusion processes, etc .

The reasons for the variation of the thermovoltage at step edges are not yet understood in detail. A purely geometric effect can be excluded because the sign is different for silver on the one hand and copper and gold on the other.

The high resolution scan shows that it is limited to the range of the transition between the upper and the lower terrace as observed in the topography. It may be suggested that the variation is caused by the different distribution of k-vectors of the tunneling electrons on the flat terrace and at an intermediate position of the tip between two terraces. This effect may be particularly important for the (111) surfaces of noble metals since they have no electron density at the Fermi level for $k_\parallel = 0$. The highest energy of the projected bulk states for $k_\parallel = 0$ is 0.3 eV below the Fermi level for silver, 0.9 eV for gold, and 0.82 eV for copper. The lowest value for the component of the k-vector parallel to the surface is $k_\parallel = 0.13\,\text{Å}^{-1}$ for silver, $k_\parallel = 0.17\,\text{Å}^{-1}$ for gold, and $k_\parallel = 0.222\,\text{Å}^{-1}$ for copper. All data have been taken from the work of *Kevan* and *Gaylord* [3.12]. These values might indicate why copper and gold show a similar behavior in contrast to silver. At the transition between the two terraces the tunneling electrons will have a significant $k_\parallel$. The quasi-gap of the (111) surfaces will not appear. Hence, the energy distribution of electronic states contributing to the tunneling process will be different, leading to a variation of the thermovoltage. In the formula given above this would enter by the appropriate choice of the density of states for the sample.

One might also suggest that the electronic surface states might be the cause of the observed lateral variation of the thermovoltage since it is known that step edges represent a barrier for these states. As will become evident in the following a variation of the electronic density of states at the surface can be observed by the thermovoltage, however, this does not explain the relatively strong signal at the step edges.

The data in Fig. 3.8 reveal a tiny modulation in the thermovoltage according to the atomic structure. To analyze this in more detail the (111) surfaces of silver, copper and gold have been investigated at high resolution. Figure 3.9 shows the corresponding topography and thermovoltage for Au(111). The corrugation in the topography is about 0.01 nm: the modulation of the thermovoltage amounts to 5 μV. Furthermore the underlying $22 \times \sqrt{3}$ reconstruction can be recognized. The domain boundaries are running from the lower left to the upper right. The example clearly demonstrates that the thermovoltage can be measured with a lateral resolution which is comparable to the one of an STM. To exclude experimental artefacts several tests have been performed. For example, the variation of the thermovoltage remains unchanged if the STM is operated in the so called "constant height" mode, hence it is independent of the motion of the tip normal to the surface which exhibits the atomic periodicity in the case of the "constant conductance" mode. As expected the signal in the thermovoltage vanishes while the corrugation in the topography stays, if there is no temperature difference between tip and sample.

The influence of surface states on the thermovoltage may be seen in Fig. 3.10. It shows an area of about 60×70 nm of a Au(111) surface. Figure 3.10a represents the topography. To enhance the visualization of small fea-

(a)

(b)

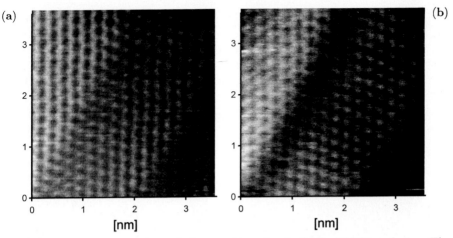

Fig. 3.9a, b. Topography and thermovoltage for Au(111) at high resolution. The scan area is about $5 \times 5\,\text{nm}$, the tunneling conductance is held constant at about $10^{-5}\,\text{A/V}$. (a) topography (gray scale representation) (b) thermovoltage (gray scale representation)

tures a shading effect has been superposed to the gray scale representation. As true geometric structures several steps and the $22 \times \sqrt{3}$ reconstruction can be clearly recognized. A closer look reveals a nontopographic effect at the step edges. The observation resemble the findings on Cu(111) by *Crommie* et al. [3.13] and on Au(111) by *Hasegawa* et al. [3.14]. On the upper terrace about two oscillations are visible in the contour of constant conductance due to the modulation of the local density of states caused by the standing waves of the surface state. Experimentally a periodicity of 1.8 nm is observed in agreement with the wave vector at the Fermi level. This is to be expected since without bias the electrons tunneling through the vacuum barrier have an energy close to the Fermi-level within a few kT. The dependence of the LDOS as function of the distance to the step is given by the zeroth order Bessel function [3.15] if one neglects the effect of thermal broadening. Due to the rapid decay only the first oscillations are visible.

The interference fringes can be much better recognized in thermovoltage which has been measured simultaneously. The data are displayed in Fig. 3.10b. Apparently the interference of the surface state not only happens at the step edges but also at local defects.

To understand the correlation between the effect of the surface state on the topography and the thermovoltage observed by STM, a line scan normal to a step edge is displayed in Fig. 3.11. To avoid complications by additional effects from the reconstruction the step edge in the middle of Fig. 3.10 has been chosen where reconstruction is normal to the step (this not visible in this image because the shading illuminates parallel to the rows of the reconstruction but it is indicated by the kinks in the step). Curve a shows the

(a)

(b)

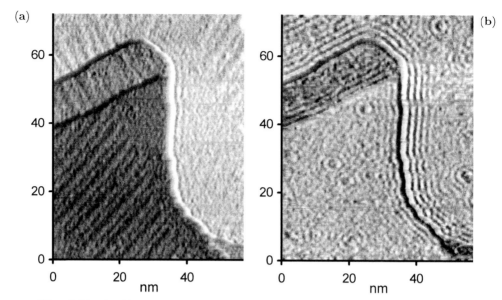

Fig. 3.10a, b. Topography and thermovoltage for a Au(111) surface. The scan area is $60 \times 70\,\mathrm{nm}^2$, the tunneling conductance is held constant at about $10^{-8}\,\mathrm{A/V}$. (a) topography (superposed gray scale and shaded representation); (b) thermovoltage (superposed gray scale and shaded representation)

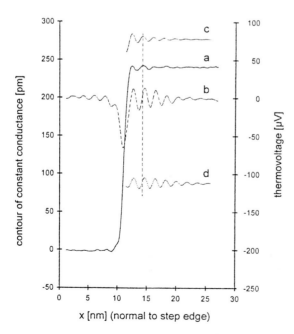

Fig. 3.11. Cross section through the interference pattern normal to a step edge: (a) topography of a section of a single line of Fig. 3.10 (contour of constant tunneling conductance); (b) measured thermovoltage; (c) calculated contour of constant tunneling conductance (d) calculated thermovoltage

contour of constant tunneling conductance. While the modulation due to the reflection of the surface state is clearly visible on the upper terrace it can only be guessed on the lower. The maximum on the upper terrace that is closest to the edge appears to be lowered due to the superposition to the topography of the step. The envelope of these oscillations is monotonically decreasing with growing distance from the step edge. In contrast the amplitude of the thermovoltage shown in curve b remains almost constant up to the third maximum before it starts decreasing. The maximum amplitude is $27\,\mu V$ which corresponds to a thermopower of $2.7\,\mu V/K$. The periodicity agrees well the predicted value of $1.8\,nm$ and with the one in the "topography". However, the phase is offset by approximately $\pi/2$. Comparing the amplitude of the oscillations on the lower and the upper terrace it may be concluded that the coefficient of reflection of the step edge is about five times smaller for the lower terrace.

To substantiate the assumption that the observed variation of the thermovoltage is due to the reflection of the surface state the thermovoltage has been evaluated based on the assumption that the density of states of the sample is composed of two terms:

$$\rho_S = \rho_B + \rho_{SS} , \tag{3.16}$$

where ρ_B is the density of states due to the projected bulk states, and ρ_{SS} the density of states of the surface state

Since we are only interested in the variation induced by the surface state, a simple approximation may be used for the projected bulk states which provide the background signal. As detailed values for density of states for the Au(111) surface were not available, ρ_B is approximated by the bulk values. For the energy range of several kT around $E_F = 0$ it can be well approximated by the linear function

$$\rho_B(E) = \rho(0) \left[1 + \frac{\rho'(0)}{\rho(0)E} \right] \exp(-2\kappa z) , \tag{3.17}$$

where $\rho(0) = 0.25\,eV^{-1}$ and $\rho'(0)/\rho(0) = -0.05\,eV^{-1}$ have been taken from the work of Jepsen et al. [3.16]. The exponential function provides the decay into the vacuum with an inverse decay length of

$$\kappa = (1/\hbar)\sqrt{2m(\phi - E)} , \tag{3.18}$$

with $\phi = 4.65\,eV$.

If the surface state electrons are assumed to be totally reflected at the lateral position $x = 0$ (e.g., by a step on the surface), the lateral variation has to be described by a Bessel function [3.15]

$$\rho_{SS} = g[1 - J_0(2k_\| x)] \exp(-2\alpha z) , \tag{3.19}$$

where g gives the relative weight with respect to the projected bulk states. The x-axis is assumed to be normal to the step. The surface state can be well described by a parabolic dispersion relation with an effective electron mass

$m^* = 0.28\,\mathrm{m}$ and an energy $E_S = -0.41\,\mathrm{eV}$ for the band edge [3.13], the component of the $\boldsymbol{k}$-vector parallel to the surface is

$$k_\| = (1/\hbar)\sqrt{2m^*(E - E_S)} \qquad (3.20)$$

and the decay length becomes

$$\alpha = (1/\hbar)\sqrt{2m(\phi - E) + \hbar^2 k_\|^2} \ . \qquad (3.21)$$

As little is known about the detailed density of states of the tunneling tip, we use the same values as for the projected bulk states of the sample:

$$\rho_T(E) = \rho(0)\left[1 + \frac{\rho'(0)}{\rho(0)}E\right] \ . \qquad (3.22)$$

The contribution of the surface state leads to an oscillatory behavior of the density of states as function of the energy. Since the argument of the Bessel function contains the product of $k_\|$ and x, the oscillations become more and more rapid with increasing distance x from the step. This prohibits the application of the approximation by *Støvneng* et al. [3.7] which requires a linear approximation for $\rho_S(E)$. Therefore, the integrals for V_{th} and σ have been evaluated numerically.

It is found that I_{th} is dominated by a large positive background resulting from the energy dependence of the tunneling process. The surface state leads to an additional term which is approximately proportional to the derivative of ρ_{SS} with respect to the energy, hence, $J_1(2k_\| x)$. This leads to the increase of the amplitude of the first oscillations with increasing x. If x is further increased however, the thermal broadening leads to a damping of the oscillations.

The conductance is also composed of two positive terms. The first which is due to the projected bulk states does not depend on x. The second which is given by the average over $\rho_{SS} = g[1 - J_0(2k_\| x)] \exp(2 - \alpha z)$ is also positive except at $x = 0$ where it vanishes.

Since the thermovoltage is given by the ratio $V_{th} = -I_{th}/\sigma$, a negative thermovoltage with a small modulation due to the surface state is predicted in good agreement with the experimental observation.

The qualitative behavior of the lateral variation of the thermovoltage is hardly influenced by the choice of the parameters for states of the tip, the projected bulk states or the relative weight of the surface state. However, the relative amplitude of the maxima and minima depends strongly on dE/dk, the slope of the dispersion relation of the surface state.

Curves c and d of Fig. 3.11 show the result of a numerical calculation for the contour of constant conductance and the thermovoltage. By comparison to the experimental curve the relative contribution of the surface state is determined to 4 % assuming that the surface state is totally reflected at the edge of the upper terrace. Using this parameter the thermovoltage can be unambiguously calculated. Despite the relative large offset and the experimental errors the periodicity as well as the envelope of the oscillations is well described.

The examples which have been discussed so far showed the lateral variation of the thermovoltage on homogeneous metallic surfaces. A very intriguing application of the thermovoltage in scanning tunneling microscopy is the possibility to analyze the composition of inhomogeneous samples.

As an example the growth of copper on Ag(111) has been investigated. From investigations by electron diffraction [3.17] and Auger spectroscopy [3.18], it was known that one may obtain a pure island growth (Vollmer-Weber). The sample has been prepared in two steps. First the silver substrate is prepared by evaporating about 200 nm silver at a rate of 1 nm/s onto mica kept at a temperature of 500 K. This leads to flat (111) oriented terraces. In a second step the equivalent of two monolayers of copper is evaporated onto the silver at a rate of 0.1 nm/s and a substrate temperature of 550 K.

The observed topography and thermovoltage are displayed in Fig. 3.12. The topography shows circular islands on a fairly flat substrate. It is very intuitive to conclude from the surface structure that the flat areas represent the clean silver substrate and that islands are formed by copper. However, it cannot be excluded that, e.g., the substrate may be covered by one or two monolayers of copper. The thermovoltage displayed in the lower part of the figure clearly shows that for this particular sample the intuition is correct. The dark areas correspond to a thermovoltage of $-400\,\mu V$ which is characteristic for silver. The bright areas exhibit about $-100\,\mu V$ which fits well for copper. (It should be noted that the absolute value of the thermovoltages is somewhat larger than in Fig. 3.4 because of a larger temperature difference of about

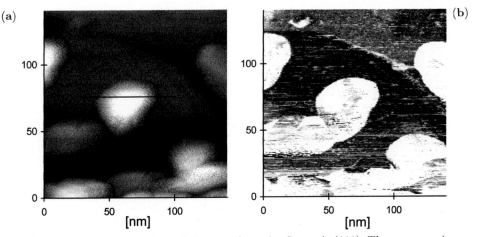

Fig. 3.12a, b. Topography and thermovoltage for Cu on Ag(111). The scan area is 140×140 nm, the tunneling conductance is held constant at about 10^{-8} A/V. The *black line* indicates the position of the cross section shown in Fig. 3.13. (a) topography varying by about 7 nm in total; (b) simultaneously measured thermovoltage ranging from -400 to $-50\,\mu V$

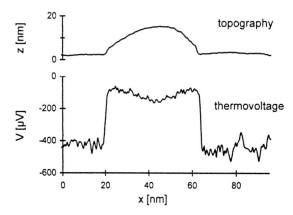

Fig. 3.13. Line scan across the central copper island of Fig. 3.12. The *upper trace* shows the topography, the *lower trace* the corresponding thermovoltage

7 K.) The contrast between silver and copper can be seen very clearly in the line scan shown in Fig. 3.13.

While the copper islands have a thickness of many atomic layers the following example of hetero-epitaxy of silver on Au(111) shows the application to monolayer islands (Fig. 3.14). The gold surface has been prepared as previously. A submonolayer of silver was evaporated onto the gold surface at room temperature. Under these conditions the single silver atoms have a high mobility and they diffuse to step edges or form larger islands. The resulting structures are not stable in long terms but it is possible to acquire images by STM, however, at some places changes are found by comparing subsequent scans.

Figure 3.14a shows the topography of the Ag/Au(111) surface. Several terraces are visible and the shape of the step edges suggest the presence of silver

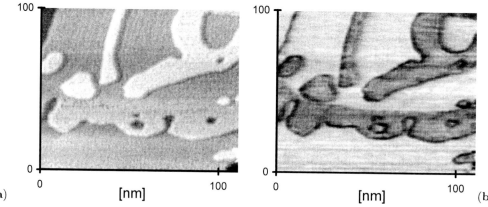

Fig. 3.14a, b. Topography and thermovoltage for a fraction of a monolayer of silver on Au(111). The scan area is $110 \times 100 \, nm^2$, the tunneling conductance is held constant at about $10^{-8} \, A/V$. (**a**) topography (gray scale representation); (**b**) thermovoltage (gray scale representation)

on the surface, e.g. the formation of holes is not found on an annnealed film of Au(111). However, based on the topography it cannot be decided to what extend the surface is covered by silver. This may be immediately analyzed by looking at the thermovoltage displayed in Fig. 3.14b, which provides a clear contrast between silver and gold. The bare gold substrate appears bright, the areas covered by a monolayer of silver are dark gray corresponding to thermovoltages of $+20\,\mu V$ and $-25\,\mu V$ respectively. The findings support the model for the diffusion of the silver proposed by *Chambliss* and *Wilson* [3.19]. After landing on the surface the atoms will rapidly diffuse on the terrace till they reach a step edge from an upper terrace. Apparently the step edges act as barriers for the diffusion. As the atoms may not step down to a lower terrace a hole may not be filled. Hence the holes are only found in areas of silver as can be immediately verified in Fig. 3.14.

For further application, it may be important that the investigation of the thermovoltage across the tunneling barrier is not restricted to purely metallic systems. As an example, a monolayer of copper-phthalocyanine (CuPc) has been investigated on Ag(111). This organic molecule has a fourfold symmetry with the copper atom in its center. From previous experiments it is known that these molecules form well ordered monolayers on Ag(111) [3.20]. Furthermore, the investigation of the current-to-voltage curve reveals that although the characteristic is not linear there is a substantial conductivity at zero bias. The latter is necessary for the proper function of the potentiometry at high resolution.

Figure 3.15 shows the topography and the thermovoltage for an oblique superstructure of CuPc on the silver substrate. The periodicity is found as well in the topography as in the thermovoltage. However, the appearance of the molecule is different. The topography shows more or less the shape of the

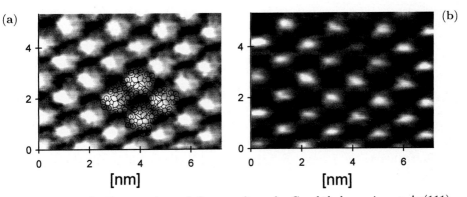

Fig. 3.15a, b. Topography and thermovoltage for Cu-phthalocyanine on Ag(111). The scan area is $7 \times 7\,nm$, the tunneling conductance is held constant at about $10^{-8}\,A/V$. (a) topography varying by about 0.3 nm with superposed molecular superstructure; (b) simultaneously measured thermovoltage ranging from -300 to $-200\,\mu V$

molecule, while the variation of the thermovoltage induced by the molecule is the strongest within a small area around the center. A detailed analysis might provide a better understanding of the contribution of the molecular states to the tunneling process.

3.4 Conclusion and Outlook

The investigation of the thermovoltage across the tunneling barrier is interesting for many reasons. First of all, it represents an important test for the theoretical description of the tunneling process of electrons across the gap of a STM. With the exception of the surface states on Au(111), the theory has only been used to provide a qualitative understanding of the experimental findings. However, a detailed knowledge of the electronic density of states should allow quantitative predictions in the near future.

On the other hand, the thermovoltage provides an additional analytic tool for STM which may be useful in many applications. For the investigation of interference phenomena of electronic surface states, it provides additional information which is not obtained by conventional tunneling spectroscopy. The variation of the thermovoltage at monoatomic steps provides an example of how it may be used to investigate the very subtle influence of surface defects on the local density of electronic states. At high lateral resolution the thermovoltage even resolves variations which correspond to the atomic periodicity of the surface, e.g., for Au(111), Ag(111) and Cu(111).

For practical application the contrast between different atomic species seems rather intriguing, e.g., for the analysis of growth processes, or alloying of different metals. Recent experiments on Au/Ag alloys show that the lateral resolution may be as good as 0.2 nm. An important question will be how many atomic layers (or how many atoms within a cluster) are needed to supply the value of the thermovoltage which is characteristic for bulk material.

It should be noted that the thermovoltage may also be used as a local thermometer [3.21] if the thermopower of the tunneling gap is known.

Acknowledgements

I would like to acknowledge the fruitful collaboration with D.H. Hoffmann, T. Kunstmann, and J. Seifritz at the University of Stuttgart, and J-Y. Grand, A. Rettenberger, K. Läuger, P. Leiderer, and K. Dransfeld at the University of Constance.

References

[3.1] C.C. Williams, H.K. Wickramasinghe: Nature **344**, 317 (1990); see also *Scanning Tunneling Microscopy II*, ed. by R. Wiesendanger, H.-J. Güntherodt, 2nd edn., Springer Ser. Surf. Sci., Vol. 28 (Springer, Berlin, Heidelberg 1995)
[3.2] M. Kohler: Ann. d. Physik **38**, 542 (1940)
[3.3] C.R. Leavens, G.C. Aers: Solid State Commun. **61**, 289 (1987)
[3.4] J.A. Støvneng, P. Lipavský: Phys. Rev. B **42**, 9214 (1990)
[3.5] J. Tersoff, D.R. Hamann: Phys. Rev. B **31**, 805 (1985)
[3.6] J. Bardeen: Phys. Rev. Lett. **6**, 57 (1961)
[3.7] J.B. Xù, K. Läuger, R. Möller, K. Dransfeld, I.H. Wilson: J. Appl. Phys. **76**, 7209 (1994)
[3.8] P. Muralt, D.W. Pohl: Appl. Phys. Lett. **48**, 514 (1986)
[3.9] J.R. Kirthley, S. Washburn, M.J. Brady: Phys. Rev. Lett. **60**, 1546 (1988)
[3.10] J.P. Pelz, R.H. Koch: Rev. Sci. Instrum. **60**, 301 (1989)
[3.11] S. Buchholz, H. Fuchs, J.P. Rabe: J. Vac. Sci. Technol. B **9**, 857 (1991)
[3.12] S.D. Kevan, R.H. Gaylord: Phys. Rev. B **36**, 5809 (1987)
[3.13] M.F. Crommie, C.P. Lutz, D.M. Eigler: Nature **524**, 363 (1993)
[3.14] Y. Hasegawa, Ph. Avouris: Phys. Rev. Lett. **71**, 1071 (1993)
[3.15] M.F. Crommie, C.P. Lutz, D.M. Eigler: Nature **363**, 524 (1993)
[3.16] O. Jepsen, D. Glotzel, A.R. Mackintosh: Phys. Rev. B **23**, 2684 (1981)
[3.17] C.T. Horng, R.W. Vook: J. Vac. Sci. Technol. **11**, 140 (1974)
[3.18] M.J. Gibson, P.J. Dobson: J. Phys. F **5**, 864 (1975)
[3.19] D.D. Chambliss, R.J. Wilson: J. Vac. Sci. Technol. B **9**, 928 (1991)
[3.20] J.Y. Grand, D. Hoffmann, T. Kunstmann, J. Seifritz, R. Möller: submitted to Surf. Sci.
[3.21] J.M.R. Weaver, L.M. Wapita, H.K. Wickramasinghe: Nature **342**, 783 (1989); see also, *Scanning Tunneling Microscopy II* ed. by R. Wiesendanger, H.-J. Güntherodt, 2nd edn., Springer Ser. Surf. Sci., Vol. 28 (Springer, Berlin, Heidelberg 1995) Chap. 6 Related Scanning Techniques by H.K. Wickramasinghe, p. 217–220

4. Spin-Polarized Scanning Tunneling Microscopy

R. Wiesendanger

With 19 Figures

Since the invention of the Scanning Tunneling Microscope (STM) [4.1] the question arose whether this technique, besides its ability to probe the Local Electronic Density of States (LDOS) [4.2], can also be made sensitive to the electron spin, thus offering the opportunity to study magnetic structures down to the atomic scale. Spin-polarized tunneling was already discovered in the seventies when planar tunnel junctions, either ferromagnet-oxide-superconductor [4.3] or ferromagnet-semiconductor-ferromagnet junctions [4.4], were studied. However, it took until 1990 for the first successful demonstration of vacuum tunneling of spin-polarized electrons with the STM [4.5].

Recent developments in the fields of thin-film magnetism and magnetic multilayer structures are demanding high-resolution magnetic imaging techniques, such as Spin-Polarized Scanning Tunneling Microscopy (SPSTM), to address the important issue of the relationship between structural and magnetic properties in systems of reduced dimension. In addition, the ability to perform spin-dependent transport measurements with an adjustable vacuum gap at high spatial resolution, as realized in SPSTM, offers novel opportunities for fundamental studies of Giant MagnetoResistance (GMR) effects and oscillatory exchange coupling as discovered in metallic multilayer structures [4.6, 7] and more recently studied in ferromagnet-semiconductor-ferromagnet trilayers [4.8], in ferromagnet-oxide-ferromagnet junctions [4.9] or in granular magnetic films [4.10, 11]. Since GMR effects have become of significant importance in magnetic sensor as well as in magnetic storage technology, it is of considerable interest to perform fundamental studies with obviously the simplest type of interlayer, i.e., vacuum.

In the following, we shall first provide a brief review of the pioneering work in the field of spin-polarized tunneling studies based on planar ferromagnet-oxide-superconductor junctions. Subsequently, we shall focus on spin-polarized tunneling between two magnetic electrodes, first in the planar junction geometry and second in the STM-type geometry. Furthermore, we shall discuss an alternative approach towards SPSTM based on electron tunneling between an optically pumped semiconductor and a ferromagnetic counter-electrode. Finally, we shall review other spin-sensitive STM experiments including the real-space observation of spin precession of paramagnetic spin centers.

4.1 Spin-Polarized Tunneling
Between a Ferromagnet and a Superconductor

Spin-polarized tunneling was first observed in planar superconductor-oxide-ferromagnetic-metal tunnel junctions [4.3, 12–14]. The experimental setup is depicted in Fig. 4.1. Usually, thin (approximately 5 nm) films of Al were deposited as narrow lines, followed by a surface oxidation treatment which resulted in insulating Al_2O_3 tunnel barriers of typically 1–2 nm thickness. Finally, a second electrode from a ferromagnetic material was then deposited as a cross strip resulting in a tunnel junction with an effective area of typically 1 mm^2. The tunneling conductance, i.e., the derivative of the tunneling current I with respect to the applied bias voltage U, was then measured as a function of U. Since the outstanding work by *Giaever* [4.15] it is well known that the dI/dU–U-characteristic of a superconductor-oxide–normal–metal tunnel junction reflects the BCS density of states of the superconductor (Fig. 4.2). *Meservey* et al. [4.12] showed furthermore that the density of states of a superconducting thin film in a strong parallel magnetic field is split into a spin-up and a spin-down part. This splitting leads to two BCS-type density-of-states curves shifted in energy by $\pm \mu H$ (μ is the electron magnetic moment, H is the applied magnetic field) with respect to the density-of-states curve in the absence of a magnetic field (Fig. 4.3).

Now consider a ferromagnetic counter-electrode where the two parts (spin up and spin down) of the spin-dependent density of states are shifted relative to each other by the exchange energy (Fig. 4.4) leading to a different density of states at the Fermi energy E_F for the two different spin directions. It is obvious that this difference in the number of spin-up electrons and spin-down electrons at E_F will lead to an asymmetry in the dI/dU–U characteristic (Fig. 4.5) which is indeed observed experimentally.

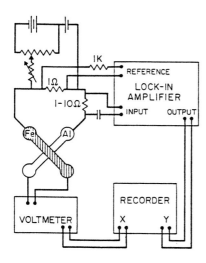

Fig. 4.1. Experimental setup for conductance vs voltage measurements on planar superconductor-oxide–ferromagnetic-metal tunnel junctions [4.13]

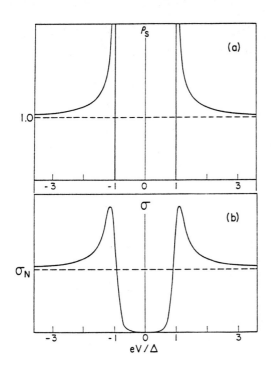

Fig. 4.2. (a) BCS density of states of a superconductor as a function of energy. (b) Normalized conductance of a superconductor-oxide–normal-metal tunnel junction [4.13]

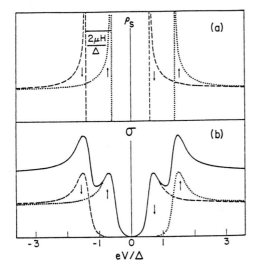

Fig. 4.3. (a) Splitting of the density of quasiparticle states into spin-up (*dotted*) and spin-down (*dashed*) parts by an applied magnetic field H. (b) Spin-up conductance (*dotted*), spin-down conductance (*dashed*), and the total conductance of a superconductor-oxide–normal-metal junction in an external magnetic field [4.13]

It is convenient to define a spin polarization P of the tunneling current by:

$$P = \frac{N_\uparrow - N_\downarrow}{N_\uparrow + N_\downarrow} , \tag{4.1}$$

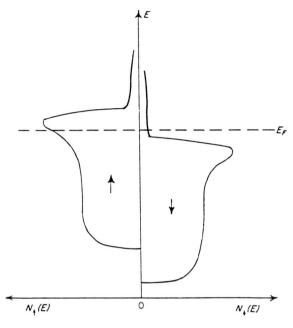

Fig. 4.4. Simple d-band model of a ferromagnet with the two parts of the spin-dependent density of states shifted relative to each other by the exchange energy leading to an unequal density of states at the Fermi energy E_F for the two different spin states

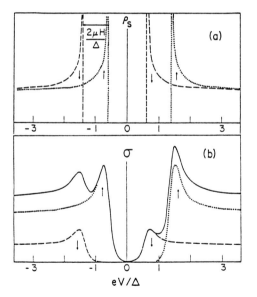

Fig. 4.5a, b. Superconductor-oxide–ferromagnetic-metal tunneling. (a) BCS density of states of a superconductor as a function of energy in a magnetic field H. (b) Normalized conductance for each spin direction (*dotted* and *dashed* curves) and the total conductance (*solid line*) for the superconductor-oxide–ferromagnetic-metal junction [4.13]

where $N_\uparrow$ ($N_\downarrow$) denotes the number of tunneling electrons with magnetic moment aligned parallel (antiparallel) to the direction of the magnetic field. By neglecting spin-orbit scattering in the superconducting electrode, which is a good first approximation for sufficiently thin films, and by assuming the absence of spin scattering in the tunnel barrier, *Tedrow* and *Meservey* [4.13]

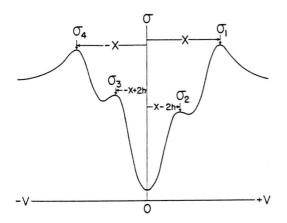

Fig. 4.6. Values of measured conductance chosen to obtain (4.2) for the polarization P [4.13]

showed that the polarization P can be determined experimentally from the relationship

$$P = \frac{(\sigma_4 - \sigma_2) - (\sigma_1 - \sigma_3)}{(\sigma_4 - \sigma_2) + (\sigma_1 - \sigma_3)} \, , \tag{4.2}$$

where σ_i are chosen values of the measured conductance defined in Fig. 4.6.

Since the tunneling current in these experiments is mainly dominated by the mobile itinerant d_i electrons of the ferromagnetic electrode with energy close to E_F which obtain their polarization through exchange interaction with the localized d_l electrons, *Tedrow* and *Meservey* measured the polarization of these d_i electrons in a thin shell within 1 meV of the Fermi surface. Table 4.1 contains a summary   of measured values of the polarization P for ferromagnetic films of Fe, Co, Ni, and Gd. The highest degree of polarization (44 %) was obtained for Fe thin films. It should be noted that photoelectrons, in contrast to tunneling electrons, are dominated by localized electrons within a much larger energy window (typically of the order of electronvolts) below the Fermi level. Therefore, values for the spin polarization as measured by spin-polarized tunneling and spin-resolved photoelectron spectroscopy [4.16] generally can not quantitatively be compared with each other [4.17].

Table 4.1. Percent polarization P measured in thin ferromagnetic films based on superconductor-oxide–ferromagnetic-metal tunneling. From [4.13]

	Tunneling [%]	Photoemission [%]
Fe	+44	+54
Co	+34	+21
Ni	+11	+15
Gd	+ 4.3	+ 5.7

The concept of using the spin-split density of states of a superconductor in a magnetic field acting as a very efficient spin-detector for the tunneling electrons from the ferromagnetic counter-electrode is undoubtedly a very elegant approach towards spin-polarized tunneling experiments. However, it has not yet successfully been applied in a STM-type geometry, mainly because of a number of instrumental requirements which are difficult to fulfill simultaneously (e.g., low-temperature, high magnetic field, ultra-high vacuum, vibration isolation, etc.). However, several different groups worldwide are working on solutions for these (technical) problems.

4.2 Spin-Polarized Tunneling Between Two Magnetic Electrodes

4.2.1 Planar Tunnel Junctions

The experiments based on planar superconductor-oxide-ferromagnet junctions have shown that the tunneling electrons are spin-polarized due to a different density of states of majority and minority spin electrons at E_F of the ferromagnetic counter-electrode. If we now consider electron tunneling between two ferromagnetic electrodes, we can expect that the tunneling conductance depends on the magnetization state of the electrodes. This was first observed by *Julliere* [4.4] in planar ferromagnet-semiconductor-ferromagnet junctions and by *Maekawa* and *Gäfvert* [4.18] in planar ferromagnet-oxide-ferromagnet tunnel junctions. To explain the experimental observations, *Slonczewski* [4.19, 20] considered a one-dimensional tunnel junction with a rectangular nonmagnetic barrier separating two homogeneously magnetized ferromagnetic electrodes. The directions of the internal magnetic fields within the ferromagnetic electrodes were assumed to differ by an angle θ (Fig. 4.7). Based on several assumptions (conducting electrodes described as free-electron metals at zero temperature, limit of small bias voltage and limit of small barrier transmission), the following expression for the conductance s of the ferromagnet-insulator-ferromagnet junction in the case of two identical ferromagnetic electrodes was obtained:

$$\sigma = \sigma_{\text{fbf}} \left(1 + P_{\text{fb}}^2 \cos\theta\right) \ , \qquad |P_{\text{fb}}| \leq 1 \tag{4.3}$$

where P_{fb} denotes the effective spin polarization of the ferromagnet-barrier interface. σ_{fbf} is a mean conductance which is proportional to $\exp(-2\kappa d)$ with the barrier width d and the decay constant κ. If the two ferromagnetic electrodes are different, the conductance becomes

$$\sigma = \sigma_{\text{fbf}'} \left(1 + P_{\text{fb}} P_{\text{f}'\text{b}} \cos\theta\right) \ . \tag{4.4}$$

As directly evident from (4.4), the tunnel conductance depends on the relative orientation of the magnetization direction within the two ferromagnetic electrodes as well as on the values of the effective polarization of the ferromagnet-barrier couples.

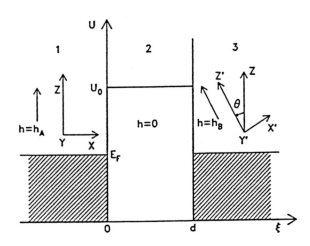

Fig. 4.7. Schematic potential energy diagram for two metallic ferromagnets separated by an insulating tunnel barrier. The directions of the internal molecular fields h_A and h_B within the ferromagnets differ by an angle θ [4.20]

For the two special cases of parallel and antiparallel alignment of the directions of the internal magnetic fields, one finds

$$\sigma_{\uparrow\uparrow} = \sigma_{\text{fbf}'} \left(1 + P_{\text{fb}} P_{\text{f}'\text{b}}\right) , \tag{4.5}$$

$$\sigma_{\uparrow\downarrow} = \sigma_{\text{fbf}'} \left(1 - P_{\text{fb}} P_{\text{f}'\text{b}}\right) , \tag{4.6}$$

Therefore, it is easily verified that

$$\frac{\sigma_{\uparrow\uparrow} - \sigma_{\uparrow\downarrow}}{\sigma_{\uparrow\uparrow} + \sigma_{\uparrow\downarrow}} = P_{\text{fb}} \cdot P_{\text{f}'\text{b}} \equiv P_{\text{fbf}'} . \tag{4.7}$$

In this equation, the combined effective polarization of the tunnel junction $P_{\text{fbf}'} = P_{\text{fb}} \cdot P_{\text{f}'\text{b}}$ has been introduced. It can experimentally be determined by measuring the ratio $(\sigma_{\uparrow\uparrow} - \sigma_{\uparrow\downarrow})/(\sigma_{\uparrow\uparrow} + \sigma_{\uparrow\downarrow})$. By using planar tunnel junctions, the value of $P_{\text{fbf}'}$ represents a spatial average over the macroscopic (ca. $1\,\text{mm}^2$) junction area. The interpretation of experimental data can be complicated by the details of the magnetic domain structure of the ferromagnetic electrodes and the magnetization reversal processes in the area of the tunnel junction [4.21].

The advantage of using tunnel junctions with ferromagnetic electrodes rather than superconductor-oxide-ferromagnet junctions is that spin-polarized tunneling can be observed at room temperature and without a strong external magnetic field. Another approach towards spin-polarized tunneling between a ferromagnet and a paramagnet is based on the consideration of the ferromagnet exchange-field penetration into the tunnel barrier region, leading to a spin selection in electron tunneling [4.22], as illustrated in Fig. 4.8. The ferromagnet exchange field penetration into the barrier region can be described by

$$B(s) = B_{\text{o}} \exp(-2\kappa s) , \tag{4.8}$$

where B_{o} is the molecular field in the bulk ferromagnet and

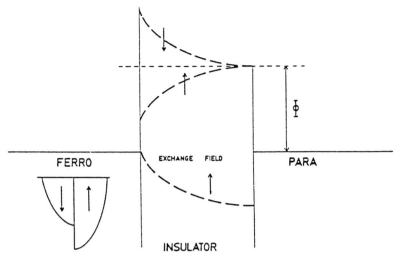

Fig. 4.8. Potential energy diagram for a ferromagnet-insulator-paramagnet tunnel junction illustrating the penetration of the ferromagnet exchange field into the barrier region and the splitting of the effective tunnel barrier height for electrons with spin-up and spin-down orientation [4.22]

$$\kappa = (2m\phi)^{1/2}/\hbar \tag{4.9}$$

with the tunnel barrier height ϕ. The penetration of the exchange field into the barrier region leads to a spin-splitting of the effective tunnel barrier height ϕ_{eff}, as seen by electrons with either spin up or spin down orientation

$$\phi_{\text{eff}} = \phi \mp \mu_B B(s) , \tag{4.10}$$

where μ_B is the Bohr magneton. Consequently, the transmission

$$T \propto \exp\left(-A\phi_{\text{eff}}^{1/2}d\right) \tag{4.11}$$

with $A = 2(2m)^{1/2}/\hbar$, and d being the width of the tunnel barrier, will be enhanced for spin-up (majority) electrons and depressed for spin-down (minority) electrons, leading to a finite spin polarization of the tunneling current.

Spin-polarized tunneling can be observed even in the case of two normal-metal electrodes in zero applied field if a ferromagnetic barrier, e.g., EuS, is used [4.23, 24]. EuS is a ferromagnetic semiconductor with a Curie temperature of $T_c = 16.6\,\text{K}$ and a bulk band gap of $1.65\,\text{eV}$. Below T_c, the conduction band is split by the ferromagnetic exchange interaction ($\Delta E_{\text{ex}} = 0.36\,\text{eV}$ at $4\,\text{K}$). Therefore, if EuS is used as a barrier in tunnel junctions, the barrier height will be split for the two spin directions below T_c (Fig. 4.9):

$$\phi_{\uparrow,\downarrow} = \phi_\circ \mp \Delta E_{\text{ex}}(T)/2 , \tag{4.12}$$

where $\phi_\circ$ denotes the average barrier height above T_c. As a consequence, the probability of tunneling for spin-up electrons is much higher than for spin-

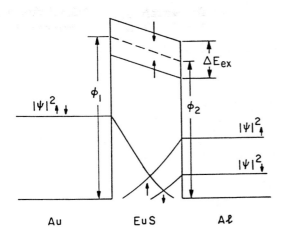

$|\Psi|^2_{\uparrow\downarrow}$

ϕ_1

ΔE_{ex}

ϕ_2

$|\Psi|^2_{\uparrow}$

$|\Psi|^2_{\downarrow}$

Au   EuS   Aℓ

Fig. 4.9. The spin-filter model for a ferromagnetic EuS tunnel barrier. The *dashed line* represents the tunnel barrier height at temperatures above the Curie temperature T_c, and the *solid lines* the tunnel barrier heights for spin-up and spin-down electrons (as indicated by the arrows) for $T \ll T_c$. The tunneling probabilities for each spin-polarization are shown schematically as the overlap of the wave functions [4.24]

down electrons because the spin-up electrons see a barrier height $\phi_\uparrow$ which is lower by ΔE_{ex} compared with the barrier height $\phi_\downarrow$ seen by the spin-down electrons. A tunnel barrier consisting of a ferromagnetic semiconductor can therefore act as a very effective 'spin filter' below T_c. In fact, spin polarization values as high as 85 % have been measured in normal-metal-EuS-normal-metal tunnel junctions [4.24]. If one of the metal electrodes is a superconductor, e.g., Al, the ferromagnetic ordering of EuS below T_c additionally leads to an exchange-induced spin-splitting of the quasiparticle density of states of the superconducting electrode even in zero external magnetic field. The spin-filter effect of EuS barriers can be used to provide a low-energy spin-polarized electron source whereas the spin-split density of states in the superconductor provides a spin-selective detector for tunneling electrons without any external magnetic field in this case. Field-emission studies of EuS-coated tungsten tips also showed a high degree of spin polarization of the field-emitted electrons below T_c [4.25, 26].

4.2.2 STM-Type Junctions with a Vacuum Barrier

The pioneering work in the field of spin-polarized tunneling based either on planar superconductor-oxide-ferromagnet [4.3] or ferromagnet-(semiconductor/oxide)-ferromagnet [4.4, 18] junctions (Fig. 4.10a, b) has laid the foundations for understanding spin-dependent transport through tunnel barriers. An independent development in the field of spin-dependent transport in metallic multilayer structures, consisting of ferromagnetic thin films separated by normal-metal or antiferromagnetic spacer layers (Fig. 4.10c), has started mid of the eighties. This has led to the discovery of GMR effects and oscillatory exchange coupling as a function of spacer layer thickness [4.6, 7]. More recently, exchange coupling has been studied in ferromagnet-semiconductor-ferromagnet structures [4.8] as well as in ferromagnet-oxide-ferromagnet structures [4.27, 28] which leads back to the earlier work in the

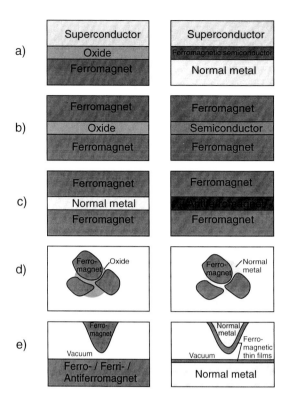

Fig. 4.10a–e. Different geometries for spin-dependent transport studies: (**a**) Planar superconductor-oxide-ferromagnet [4.3] and superconductor-ferromagnetic semiconductor–normal metal [4.23] tunnel junctions. (**b**) Planar ferromagnet–oxide–ferromagnet [4.18] and ferromagnet–semiconductor–ferromagnet [4.4, 8] junctions. (**c**) Planar ferromagnet–normal metal–ferromagnet and ferromagnet–antiferromagnet–ferromagnet [4.6] heterostructures. (**d**) Granular ferromagnetic materials in oxide [4.29, 30] or normal metal matrices. (**e**) Ferromagnet–vacuum–magnet [4.5, 50] and ferromagnetic thin-film–vacuum–ferromagnetic thin-film [4.64] tunnel junctions

field of spin-polarized tunneling by *Julliere* [4.4] and *Maekawa* and *Gäfvert* [4.18]. The MagnetoResistance (MR) measured in this case is called Tunneling MagnetoResistance (TMR). Insulating spacer layers are prefered nowadays for applications of the MR effect in magnetic field sensors because the exchange coupling between the ferromagnetic layers becomes reduced compared with the case of normal-metal spacer layers, resulting in low power devices with high field sensitivity due to low saturation fields. MR effects were even observed in granular magnetic films [4.29, 30] where spin-dependent tunneling across insulating grain boundaries acting as tunnel barriers (Fig. 4.10d) has to be taken into account to explain the experimental results quantitatively [4.31, 32].

Obviously, the most simple type of spacer layer from a conceptual point of view would be vacuum. This idea directly leads to an STM-type geometry, as illustrated in Fig. 4.10e. Vacuum tunneling of spin-polarized electrons with the STM was first observed in 1990 [4.5]. The STM geometry has three major attractive features:

1) Spin-dependent transport measurements can be performed with an adjustable vacuum gap by controlling the tip-sample separation (z-direction).

2) Spatially-resolved studies of exchange-coupling effects are possible by scanning or lateral positioning of the tip relative to the sample (in the xy-plane).
3) The relationship between topographic, electronic and magnetic structure can be studied at the atomic level by combining STM, tunneling spectroscopy and SPSTM.

4.2.2.1 Choice of Magnetic Probe Tips. Different types of magnetic probe tips can be used:

– tips from bulk magnetic material (Fig. 4.10e, left part)
– nonmagnetic tips covered with a thin film of magnetic material (Fig. 4.10e, right part)
– nonmagnetic tips with a cluster of magnetic material at the front end.

Probably the most important issue is the appropriate choice of magnetic tip material for SPSTM studies. Fe is known to exhibit a relatively high degree of spin polarization of the electronic states near the Fermi level and has indeed shown the maximum spin polarization among the ferromagnetic $3d$ transition metals in tunneling experiments performed with planar junctions [Ref. 4.13, Table 4.1]. An even better choice may be EuS-coated tungsten tips. The EuS layer acts as an efficient spin filter for electrons from the tungsten substrate tip leading to a high degree of spin polarization (Fig. 4.9). However, the Curie temperature of EuS is about $16 \, \text{K}$ and therefore low temperatures are required in this case.

Optimum spin filters at room temperature are provided by a class of materials known as half-metallic magnets [4.33–35]. Half-metallic magnets are defined as materials exhibiting metallic behavior for one spin direction, while being insulators (or semiconductors) for the opposite spin direction. As a consequence, a $100 \, \%$ spin polarization of the electronic states at the Fermi level is expected, which is highly favorable for SPSTM experiments. Several Mn-based Heusler alloys (XMnSb, where $X = $ Ni, Co, Pt) [4.33] as well as CrO_2 [4.36, 37], Fe_3O_4 [4.38, 39], etc. were theoretically predicted to be half-metallic magnets.

In principle, antiferromagnetic tip materials would be preferable to exclude the disturbing influence of the tip magnetic stray field on the magnetic structure of the sample [4.40]. The optimum choice would probably be the new class of half-metallic antiferromagnetic materials [4.34, 35].

4.2.2.2 SPSTM Experiments with Ferromagnetic CrO_2 Thin Film Tips and Cr(001) Sample. Vacuum tunneling of spin-polarized electrons in SPSTM experiments with two magnetic electrodes in zero external magnetic field was first demonstrated by using a ferromagnetic CrO_2 thin film tip and a Cr(001) sample [4.5, 41]. CrO_2 was chosen as tip material because of its outstanding properties as a conjectured half-metallic ferromagnet [4.36]. A high degree of spin polarization close to $100 \, \%$ of the electronic states near

2 eV below the Fermi level was indeed found experimentally by spin-resolved photoelectron spectroscopy [4.42]. On the other hand, a low density of states at E_F was observed in these experiments in contradiction to the conjectured half-metallic behavior of CrO_2. However, more recent optical reflectivity measurements [4.43], as well as band-structure calculations [4.37], provide strong evidence that the half-metallic nature of CrO_2 persists even up to the surface. The CrO_2 thin-film tips were prepared according to the following procedure: A CrO_2 film was deposited onto a heated Si(111) substrate covered with a 35 nm thick TiO_2 nucleation layer by decomposing CrO_3 vapor in a convection reactor. The CrO_2 thin films exhibited a large remanent in-plane magnetization. Subsequently, the Si(111) substrates were scribed and cleaved after the deposition of the CrO_2 films to obtain the shape of a tip. Finally, the Si substrate at the front end of each tip was etched back in a HF–HNO_3 solution. As a result, an overhanging CrO_2 thin-film tip was obtained which was cleaved in-situ under Ultra-High Vacuum (UHV) conditions before its use as a magnetic probe for SPSTM studies.

A Cr(001) sample was selected in this study because the topological antiferromagnetic order of this surface [4.44] provided an interesting model system for SPSTM experiments on the nanometer scale. An electrochemically polished single crystal of chromium was prepared in UHV by repeated cycles of Ar^+ ion etching and thermal annealing over a time period of several months. The successful preparation of a clean and well ordered Cr(001) surface was checked by Auger electron spectroscopy (AES), Low Energy Electron Diffraction (LEED), and STM. The subsequent use of a CrO_2 probe tip allowed to probe the alternating magnetization of the terraces separated by monoatomic surface steps (Fig. 4.11). In this case, the contribution from spin-polarized tunneling showed up by a periodic alternation of measured monoatomic step-height values (Fig. 4.12) which was not observed in the con-

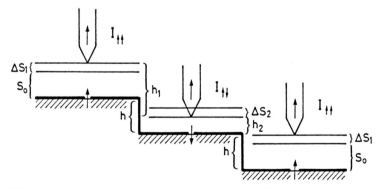

Fig. 4.11. Schematic drawing of a ferromagnetic tip scanning over alternately magnetized terraces separated by monoatomic steps of height h. An additional contribution from spin-polarized tunneling leads to alternating step heights $h_1 = h + \Delta s_1 + \Delta s_2$ and $h_2 = h - \Delta s_1 - \Delta s_2$ [4.5]

(a)   **(b)**

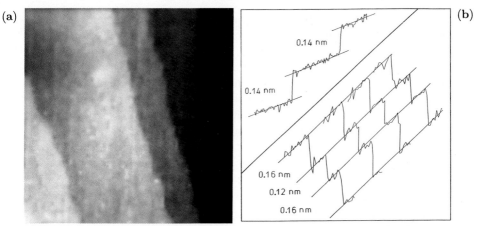

Fig. 4.12. (a) Constant-current STM image (32 nm × 32 nm) of a Cr(001) surface obtained with a CrO_2 tip, showing terraces separated by monoatomic steps with different gray levels for each terrace. (b) Single-line scans over the same three monoatomic steps taken from the STM image shown in (a). The same alternation of step-height values (0.16 nm, 0.12 nm, and again 0.16 nm) is evident in all single-line scans. The line scans are 22 nm long. *Inset*: for comparison, a single-line scan over two monoatomic steps as measured with a tungsten tip is presented. In this case, the measured step-height value (0.14 nm) is constant and corresponds to the topographic monoatomic step height. This line scan is 70 nm long [4.5]

trol experiments performed with nonmagnetic tungsten tips. The alternation of the monoatomic step-height values as measured with a CrO_2 tip was reproducibly observed in independent line sections as well as in subsequent SPSTM images [4.45]. A relationship between the deviation Δs of the measured step-height value from the topographic step-height value and the local effective spin polarization $P_{\text{fbf}'}$ of the tunnel junction was derived [4.5, 46]:

$$P_{\text{fbf}'} = \frac{\exp\left(A\sqrt{\phi}\,\Delta s\right) - 1}{\exp\left(A\sqrt{\phi}\,\Delta s\right) + 1}, \tag{4.13}$$

where $A \approx 1\,\text{eV}^{-1/2}\text{Å}^{-1}$ and ϕ denotes the local tunnel barrier height. As a remarkable consequence, the measurement of the spin polarization $P_{\text{fbf}'}$ has been reduced to a measurement of geometrical quantities ($\Delta s = \Delta s_1 + \Delta s_2$, see Fig. 4.11). The experimentally determined value of $P_{\text{fbf}'}$ at a sample bias voltage of $U = +2.5\,\text{V}$ and a tunneling current of 1 nA was found to be $(20 \pm 10)\%$. This experimentally determined value of $P_{\text{fbf}'}$ ($U = +2.5\,\text{V}$) appeared to be in reasonable agreement with a model calculation of $P_{fbf'}(U)$ for a CrO_2–vacuum–Cr(001) tunnel junction [4.45] based on the spin-dependent electronic structure for the two magnetic electrodes and the assumption of absence of spin-flip tunneling processes. The experimental observation of the dependence of $P_{\text{fbf}'}$ on the applied sample bias voltage was qualitatively in agreement with the theoretical calculation. Furthermore, it was shown that the CrO_2–vacuum–Cr(001) junction represents an interesting case in which

$P_{\mathrm{fbf'}}(U)$ switches its sign several times as a function of U [4.45]. Consequently, particular values of the applied bias voltage exist for which $P_{\mathrm{fbf'}}(U) = 0$. For such cases, the important problem of separating topographic and magnetic information can be solved easily: Pure topographic information is obtained by selecting a sample bias voltage for which $P(U) = 0$, whereas maximum magnetic contrast is obtained by choosing the sample bias voltage for which the value of $|P(U)|$ is highest. By simultaneously recording images at those two bias voltages, topographic and magnetic information can be separated.

The Cr(001) experiment constituted the first step towards STM with spin-polarized electrons [4.47] although in-plane atomic resolution was not achieved at that time, presumably due to the limited sharpness of the CrO_2 probe tips.

4.2.2.3 SPSTM Experiments with Ferromagnetic Fe Tips and Fe_3O_4(001) Sample.

For the second SPSTM experiment [4.48–50] another conjectured half-metallic magnet, Fe_3O_4, was chosen, this time as sample. To be able to reach in-plane atomic resolution, Fe tips were prepared according to the following procedure: A 0.25 mm diameter wire was first electrochemically etched until a constriction of 20–50 µm diameter was formed. The top part was then fixed in the tip holder of the STM while the bottom part was fixed in another tip holder on top of a standard sample holder. The two parts were finally pulled apart in UHV at a background pressure in the low 10^{-11} mbar regime. The microtips formed at the end of the tips reproducibly had a radius of curvature of 10 nm or even less [4.51, 52]. The formation of microtips with such a small radius of curvature did not depend on the diameter of the original constriction. Most important, the cleanness of the tips at the sharp end was limited only by the amount of bulk impurities of the wire material. Using these in-situ prepared Fe tips in-plane atomic resolution was routinely obtained on Si test surfaces as well as on the Fe_3O_4(001) surface [4.51–54].

The preparation of a clean (001) surface of a mechanically polished Fe_3O_4 natural single crystal was accomplished by in-situ thermal annealing which led to an evaporation of hydrocarbon contamination species from the surface [4.50]. It was, however, important to keep the background pressure during the thermal annealing treatment in the low 10^{-10} mbar regime at least. Furthermore, variation of the surface preparation procedure was found to have drastic influences on the surface structure of Fe_3O_4(001) [4.55].

The SPSTM study concentrated on the particular (001) plane which contains oxygen and octahedrally coordinated Fe sites (Fig. 4.13). These octahedrally coordinated Fe sites are half occupied by Fe^{3+} with an electronic and spin configuration $3d^{5\uparrow}$, and half occupied by Fe^{2+} with an additional sixth $3d$ electron of opposite spin alignment: $3d^{5\uparrow}3d^{5\downarrow}$. The additional electrons originating from the Fe^{2+} sites are responsible for the metallic conductivity of Fe_3O_4 in the bulk at room temperature because they can hop rapidly among the octahedrally coordinated sites. At the Verwey transition tempera-

(a)

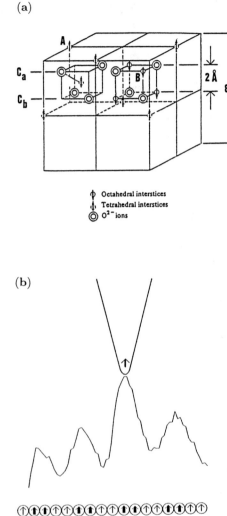

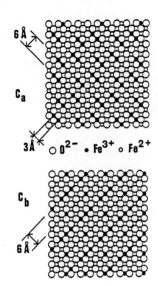

Octahedral interstices
Tetrahedral interstices
O^{2-} ions

$\circ\, O^{2-}$ $\bullet\, Fe^{3+}$ $\circ\, Fe^{2+}$

(b)

Fig. 4.13. (a) Structure of the (001) plane of Fe_3O_4 containing oxygen and octahedrally coordinated iron sites. The ordering of Fe^{3+} and Fe^{2+} on these sites in the low temperature phase of magnetite below the Verwey transition has been deduced from NMR data [4.56–58]. (b) Schematic drawing of a ferromagnetic probe tip scanning along the row of octahedrally coordinated Fe sites within the Fe–O (001) plane of magnetite

ture ($T_V \approx 120\,\mathrm{K}$ in the bulk) a metal–insulator transition sets in, leading to an ionic crystal state in which a periodicity of $12\,\text{Å}$ between Fe^{3+} and Fe^{2+} sites along the [110] direction exists (Fig. 4.13). At the (001) surface of Fe_3O_4 the Verwey transition temperature appears to be enhanced above room temperature due to the reduced coordination of surface Fe sites [4.59, 60]. This allowed to observe the $12\,\text{Å}$ periodicity between the different ionic sites in a room temperature SPSTM experiment (Fig. 4.14). The peak-to-peak corrugation of the $12\,\text{Å}$ periodic line sections as measured with Fe probe tips was typically on the order of several tenth Ångstrøm. In contrast, the $12\,\text{Å}$ periodicity was hidden in the noise level being below $0.1\,\text{Å}$ when the mea-

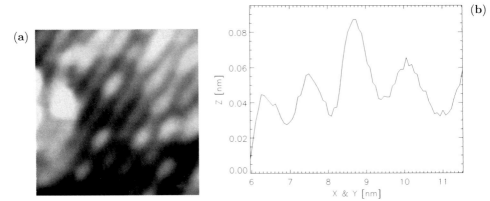

Fig. 4.14. (a) Constant-current STM image of the Fe–O (001) plane of magnetite obtained with an Fe probe tip. The rows of octahedrally coordinated Fe sites of spacing 6 Å and with modulation along these rows can clearly be seen. A period of 12 Å is preferentially observed along the Fe rows. This is particularly evident from single-line sections along the Fe rows as shown in (b). The period of 12 Å corresponds to the repeat pattern of Fe^{2+} and Fe^{3+} in the Fe–O (001) plane of magnetite (Fig. 4.13)

surements were performed with nonmagnetic tungsten tips. The enhancement of the corrugation at the 12 Å periodicity reflecting the alternation between Fe^{3+} and Fe^{2+} sites, as measured with Fe probe tips, was attributed to spin-polarized tunneling, whereas the lack of a measurable corrugation at the 12 Å periodicity, as observed with nonmagnetic tungsten tips, indicated a smooth spin-averaged density-of-states corrugation along the rows of octahedrally co-ordinated Fe sites in the [110] direction. A theoretical treatment of SPSTM by *Molotkov* [4.61] predicted that the spin-dependent part of the tunneling current is proportional to the scalar product of the local magnetization at the tip apex and crystal surface. Since the local magnetic moments of Fe^{3+} and Fe^{2+} in the (001) plane of Fe_3O_4 are parallel orientated, it must be the difference in the absolute value of the magnetic moment between Fe^{3+} and Fe^{2+} being responsible for the contrast enhancement by spin-dependent tunneling from a ferromagnetic probe tip. More recently, similar conclusions have been drawn from SPSTM experiments on (001) surfaces of artificially grown Fe_3O_4 single crystals by *Koltun* [4.62]. Again, the 12 Å periodicity between Fe^{3+} and Fe^{2+} sites could only be resolved when in-situ prepared Fe probe tips were used.

The $Fe_3O_4(001)$ experiment proved for the first time that ångstrøm-scale resolution can be obtained with a magnetic tip scanning over a magnetic sample. This opens up a new field of imaging, probing, and modifying magnetic surfaces down to the atomic level [4.63].

4.2.2.4 SPSTM Experiments with Ferromagnetic Fe Thin Film Tips and Fe Thin Film Sample. To avoid the possible influence of the magnetic stray field from a bulk ferromagnetic tip on the magnetic structure of the sample, ferromagnetic thin film tips (Fig. 4.10e, right part) are a better choice. Such probe tips again have to be prepared in-situ by evaporating a thin ferromagnetic layer onto a pre-cleaned normal-metal tip, e.g., a W tip. We recently applied such ferromagnetic Fe thin-film tips to investigate the local electronic and magnetic structure of Fe thin films on W(110) substrates [4.64]. To exclude spectroscopic differences of nonmagnetic origin between different tip materials, we first measured tunneling spectra on a non-magnetic W(110) substrate (Fig. 4.15a) both with W- and with Fe-coated W-tips. Figure 4.15b, c show two spectra obtained with an electrochemically etched W tip (Fig. 4.15b) and an in-situ Fe-coated tip (Fig. 4.15c). It is remarkable how similar these two spectra are despite the fact that the microstructure of the tips might be completely different. There is only a small shift of the dI/dU–U characteristic by about $0.3\,\mathrm{eV}$ visible between the two spectra which can easily be explained by the work function difference between a W- and a Fe-tip. Similar results were obtained with other W- and Fe-tips as well. Subsequently, a comparative spectroscopic study of two-monolayer-high Fe-islands (Fig. 4.16a) was performed. Such Fe islands are ferromagnetic at room temperature as known from torsion oscillation magnetometry measurements [4.65]. Figure 4.16b shows the typical tunneling spectrum obtained by using a W-tip whereas Fig. 4.16c reveals the corresponding spectrum obtained with Fe-coated probe tips. Apparently, drastic differences in the two spectra can be observed. The peak at about $U = +0.3\,\mathrm{V}$ in spectrum (b) can be ascribed to a d_{z^2}-like electronic state of ultra-thin Fe-films on W(110) [4.66] which was predicted earlier based on band structure calculations [4.67]. It was additionally predicted [4.67] that this electronic state should be highly spin-polarized because it results almost exclusively from (minority) spin electrons.

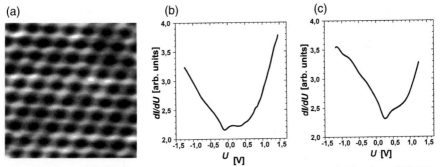

Fig. 4.15. (a) Atomically resolved STM image ($2.1\,\mathrm{nm} \times 2.1\,\mathrm{nm}$) of a clean W(110) surface. (b) Tunneling spectrum obtained with an electrochemically etched W tip on the clean W(110) surface. (c) Tunneling spectrum obtained with an in-situ Fe-coated W tip on the clean W(110) surface

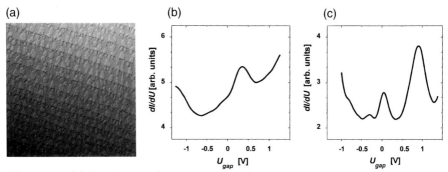

Fig. 4.16. (a) STM image (335 nm × 335 nm) showing two-monolayer-high Fe islands on an Fe-covered W(110) surface. (b) Tunneling spectrum obtained with an electrochemically etched W tip on a two-monolayer-high Fe island. (c) Tunneling spectrum obtained with an Fe-coated W tip on a two-monolayer-high Fe island

If the tunneling experiment is repeated with an Fe-coated tip (Fig. 4.16c), a minimum in the dI/dU–U characteristic at $U = +0.3$ V appears while two prominent peaks show up close to E_F and 0.8 eV. The drastic change in the tunneling spectrum might be explained by the magnetic exchange interaction across the tunnel barrier in case of two ferromagnetic electrodes. Earlier experimental work based on pairs of Fe films separated by a layer of carbon [4.68] indicated that the exchange coupling between the two Fe films critically depends on the thickness of the carbon spacer layer. The STM geometry used in [4.64], in principle, allows to study the exchange coupling effects as a continuous function of the separation between the two Fe electrodes. However, more experimental and theoretical work is needed to fully exploit the potential of SPSTS in which spin-polarized tunneling is studied as a function of the applied bias voltage and the separation between the two magnetic electrodes.

4.2.2.5 Spin-Dependent Tunneling Experiments with External Magnetic Field. By using an external magnetic field, the magnetization of the sample (or the tip) can be modulated periodically. Consequently, a portion of the tunneling current is predicted to oscillate at the same frequency. The advantage of this experimental procedure lies in the fact that lock-in detection techniques can be applied, resulting in an improvement of the signal-to-noise ratio. Initial experiments were performed in ambient air by using a single-crystal Ni tip and a permalloy sample which were likely to be in contact [4.69]. Values for the local effective polarization $P_{\mathrm{fbf'}}$ were derived in stationary experiments without scanning. However, in principle, the magnetic field can be modulated at a frequency well above the cut-off frequency of the feedback loop and the corresponding amplitude of the current oscillation can be recorded with a lock-in-amplifier simultaneously with the constant-current topograph. The spatially resolved lock-in signal then provides a map reflecting the spatial variations of the effective spin polarization of the tunnel junction.

4.3 Spin-Polarized Tunneling
Between an Optically Pumped Semiconductor
and a Ferromagnetic Conter-Electrode

Semiconductors, such as GaAs, which are optically pumped by circularly polarized light are well known as sources of spin-polarized electrons. A significant advantage of the application of GaAs in spin-polarized tunneling experiments would be that the polarization of the photo-excited electrons can easily be reversed by reversing the helicity of the circularly polarized light [4.70, 71]. However, several critical issues have to be addressed, for instance, whether a sufficient number of electrons can be excited into the semiconductor's conduction band to achieve a measurable tunneling current of spin-polarized electrons. Furthermore, thermal beating of the tip and sample by optical pumping leads to relative displacements of the two electrodes due to thermal expansion, which additionally affects the tunneling current signal.

Experiments based on a GaAs thin-film sample, optically pumped by circularly polarized light, and a ferromagnetic Ni tip (Fig. 4.17) indicated that spin-polarized tunneling can be distinguished from other spurious effects such as thermal expansion of tip and sample [4.72]. This was demonstrated by directly monitoring the variation of the tunneling current signal as the circular polarization of the light was modulated. In a complementary experiment (Fig. 4.18), spin-polarized tunneling from a ferromagnetic Ni tip into GaAs was observed by measuring the circular polarization of the recombination luminescence excited by electrons tunneling from the Ni tip into GaAs acting as a spin detector [4.73]. The spin polarization of the electrons extracted from

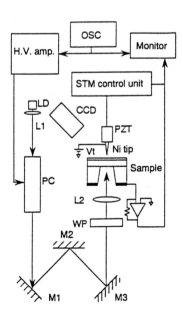

Fig. 4.17. Schematic experimental setup for a spin-polarized tunneling study based on an optically pumped semiconductor heterostructure sample and a ferromagnetic Ni tip. (LD: laser diode, L1/L2: lenses, PC: Pockels cell, M1-3: mirrors, WP: quarter-wave plate) [4.72]

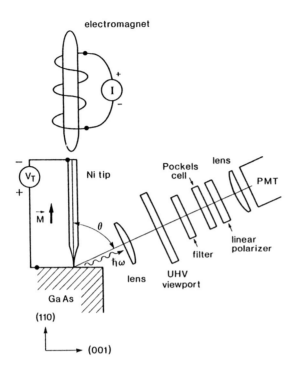

Fig. 4.18. Schematic experimental setup for a measurement of the circular polarization of the recombination luminescence excited by electrons tunneling from a ferromagnetic Ni tip into a GaAs(110) sample [4.73]

the Ni tip was found to be $P = (-31 \pm 5.6)\%$. The negative sign indicates that minority spin $3d$ electrons at the Fermi level contributed predominantly to the tunneling current in this experiment. In contrast, a positive polarization in spin-dependent tunneling between thin films of superconducting Al and ferromagnetic Ni was found earlier based on the planar junction geometry [Ref. 4.3, Table 4.1]. In this case, the tunneling current is believed to be due to s–d hybridized bands exhibiting positive polarization [4.17]. Since these bands exhibit a lower degree of polarization, the smaller value of the polarization for Ni $(P = +11\,\%)$ as measured in planar superconductor-oxide-ferromagnet junctions [4.3] can be explained. More recently, *Alvarado* [4.74] measured the degree of spin polarization of electrons tunneling from the Ni tip into a semiconductor sample at constant energy as a function of tunnel barrier width (Fig. 4.19). The results showed that the highly polarized $3d$ bands as well as the low-polarized $4sp$ bands contribute to the tunneling current and that the ratio of their tunneling probabilities depends on the barrier width and height. The tunnel barrier for the $3d$ states was estimated to be about $1\,\mathrm{eV}$ higher than for the $4sp$ states. In another experiment, *Mukasa* et al. [4.75] measured the spin polarization of electrons tunneling between a very thin GaAs sample and a ferromagnetic Ni or Fe tip as a function of the applied bias voltage.

A strong dependence of the measured spin polarization on the bias voltage is generally observed reflecting properties of the spin-dependent local electronic density of states of the tips as well as of the spin polarization of

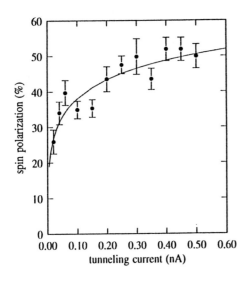

Fig. 4.19. Spin polarization of electrons tunneling from a ferromagnetic Ni tip into a $Al_{0.1}Ga_{0.9}As(110)$ sample at fixed bias voltage ($U = -1.8\,\text{V}$) as a function of the tunneling current. The variation of the tunneling current $20\,\text{pA} < I < 500\,\text{pA}$ corresponds to a change $\Delta d \approx 0.18\,\text{nm}$ of the tunnel barrier width [4.74]

excited carriers in the sample. Again, a negative spin polarization for tunneling between the GaAs sample and a Ni tip was obtained in agreement with experiments by *Alvarado* et al. [4.73]. In contrast, a positive spin polarization was measured for electron tunneling between GaAs and a ferromagnetic Fe tip.

For the application of the concept of using an optically pumped semiconductor in SPSTM, the role of sample and tip have to be changed. Recently, first experiments with an optically pumped GaAs tip and a Pt/Co multilayer sample were reported [4.76]. A circular-polarization-dependent tunneling current was found which was verified not to be due to variations of the optical power but may rather be explained by spin-polarized tunneling [4.76]. In principle, such GaAs tips can provide atomic resolution [4.77]. However, spatially resolved SPSTM data based on using an optically pumped semiconductor have not yet been reported.

4.4 Other Spin-Sensitive STM Experiments

Spin-sensitive STM experiments may be performed with nonmagnetic tips as well. For instance, the precession of individual paramagnetic spins on a partially oxidized Si(111) surface in a constant magnetic field induces a modulation in the tunneling current at the Larmor frequency [4.78, 79]. The measured radio-frequency signal from the individual paramagnetic spin center was shown to be localized over distances less than 1 nm. In contrast, ESR measures the precession of an induced macroscopic magnetization in an external magnetic field which is a result of the alignment of many (at least 10^{10}) electron spins. Valuable spectroscopic information about the properties of individual spins is lost as a result of inhomogeneous line broadening.

It was proposed that spin precession does not provide a direct contribution to the tunneling current at the Larmor frequency, but induces a time correlation between tunneling electrons with opposite spins due to interaction of the localized spin with the spins of the delocalized electrons, which would explain the experimental observations [4.80, 81]. Mechanisms based on the spin-orbit interaction were discussed controversially [4.82, 83].

4.5 Concluding Remarks

SPSTM allows to achieve the ultimate goal of polarized microscopy, to image surface spin configurations down to the atomic level. In addition, SPSTM and SPSTS can yield a variety of information about the magnetic structure of solid surfaces:

1) The effective spin polarization of the tunnel junction can be measured locally $[P(x_0, y_0)]$ as well as spatially resolved $[P(x, y)]$.
2) Distance-dependent measurements of the effective polarization $P(z)$ may provide access to the different decay of s-, p- and d-type electronic states which contribute differently to the effective polarization.
3) Bias-dependent measurements of the effective spin polarization $P(U)$ provide access to the local and spatially resolved spin-dependent electronic structure. The experimentally obtained information can be compared with energy-sliced spin density calculations for the corresponding magnetic surfaces [4.84].
4) SPSTM experiments performed with magnetic probe tips, for which the magnetization direction can be changed in an external magnetic field, may allow for an independent determination of the in-plane and out-of-plane components of the sample surface magnetization.

For further development of the highly promising SPSTM technique, considerable experimental as well as theoretical work remains to be done. On the experimental side, the most critical issue is the appropriate preparation and characterization of ferromagnetic or GaAs probe tips. What is the degree of spin polarization of the tunneling electrons from ferromagnetic or GaAs tips? What is the nanomagnetic structure at the front end of a ferromagnetic tip? Which direction of magnetization is found in the magnetic domain at the front end of such a tip? Through which spin-dependent electronic states does tunneling occur (as a function of bias and tip-sample separation)? What is the minimum size of a magnetic cluster at the front end of a nonmagnetic tip required for the observation of spin-polarized tunneling? Clearly, these questions are of great importance and also require appropriate theoretical treatment. A first theoretical study of the interaction between a magnetic nanotip and a magnetic surface was recently reported [4.85]. A strong enhancement of the magnetic moments was found on the tip's surfaces and tip's apex atoms which, however, decreases for decreasing tip-sample separation. A large antiferromagnetic coupling between a Fe tip and a Cr surface was predicted, whereas a small ferromagnetic coupling should exist between a Fe tip and a Ni surface. Though this pioneering theoretical study addressed the issue of magnetic coupling energy for tip-sample systems, spin-dependent tunneling between magnetic electrodes in a STM-type geometry has not been theoretically addressed yet. Further progress in SPSTM, however, will critically depend on an improved understanding of spin-dependent transport through a tunnel barrier in a local probe geometry.

Acknowledgements

The author would like to thank his coworkers in the field of spin-polarized scanning tunneling microscopy for their contributions. He also thanks S.F. Alvarado, A. Hernando, R. Meservey, K. Mukasa, and J.C. Slonczewski for providing figures for this chapter.

References

[4.1]   G. Binnig, H. Rohrer, Ch. Gerber, E. Weibel: Phys. Rev. Lett. **49**, 57 (1982)
[4.2]   J. Tersoff, D.R. Hamann: Phys. Rev. Lett. **50**, 1988 (1983)
[4.3]   P.M. Tedrow, R. Meservey: Phys. Rev. Lett. **26**, 192 (1971)
[4.4]   M. Julliere: Phys. Lett. A **54**, 225 (1975)
[4.5]   R. Wiesendanger, H.-J. Güntherodt, G. Güntherodt, R.J. Gambino, R. Ruf: Phys. Rev. Lett. **65**, 247 (1990)
[4.6]   P. Grünberg, R. Schreiber, Y. Pang, M.B. Brodsky, H. Sowers: Phys. Rev. Lett. **57**, 2442 (1986)
[4.7]   M.N. Baibich, J.M. Broto, A. Fert, Nguyen Van Dau, F. Petroff, P. Etienne, G. Creuzet, A. Friederich, J. Chazelas: Phys. Rev. Lett. **61**, 2472 (1988)
[4.8]   B. Briner, M. Landolt: Phys. Rev. Lett. **73**, 340 (1994)
[4.9]   C.L. Platt, B. Dieny, A.E. Berkowitz: Appl. Phys. Lett. **69**, 2291 (1996)
[4.10] H. Fujimori, S. Mitani, S. Ohnuma: Mater. Sci. Eng. B **31**, 219 (1995)
[4.11] J. Inoue, S. Maekawa: Phys. Rev. B **53**, R 11927 (1996)
[4.12] R. Meservey, P.M. Tedrow, P. Fulde: Phys. Rev. Lett. **25**, 1270 (1970)
[4.13] P.M. Tedrow, R. Meservey: Phys. Rev. B **7**, 318 (1973)
[4.14] R. Meservey: Phys. Scr. **38**, 272 (1988)
[4.15] I. Giaever: Phys. Rev. Lett. **5**, 147 (1960)
[4.16] G. Busch, M. Campagna, P. Cotti, H.Ch. Siegmann: Phys. Rev. Lett. **22**, 597 (1969)
[4.17] M.B. Stearns: J. Magn. Magn. Mat. **5**, 167 (1977)
[4.18] S. Maekawa, U. Gäfvert: IEEE Trans., MAG-18, 707 (1982)
[4.19] J.C. Slonczewski: J. de Physique, Coll. C **8**, 1629 (1988)
[4.20] J.C. Slonczewski: Phys. Rev. B **39**, 6995 (1989)
[4.21] J. Nowak, J. Rauluszkiewicz: J. Magn. Magn. Mat. **109**, 79 (1992)
[4.22] N. Garcia, A. Hernando: Phys. Rev. B **45**, 3117 (1992)
[4.23] J.S. Moodera, X. Hao, G.A. Gibson, R. Meservey: Phys. Rev. Lett. **61**, 637 (1988)

[4.24] X. Hao, J.S. Moodera, R. Meservey: Phys. Rev. B **42**, 8235 (1990)
[4.25] N. Müller, W. Eckstein, W. Heiland, W. Zinn: Phys. Rev. Lett. **29**, 1651 (1972)
[4.26] E. Kisker, G. Baum, A.H. Mahan, W. Raith, K. Schröder: Phys. Rev. Lett. **36**, 982 (1976)
[4.27] J.S. Moodera, L.R. Kinder, J. Nowak, P. LeClair, R. Meservey: Appl. Phys. Lett. **69**, 708 (1996)
[4.28] C.L. Platt, B. Dieny, A.E. Berkowitz: Appl. Phys. Lett. **69**, 2291 (1996)
[4.29] J.S. Helman, B. Abeles: Phys. Rev. Lett. **37**, 1429 (1982)
[4.30] J.I. Gittleman, Y. Goldstein, S. Bozowski: Phys. Rev. B **5**, 3609 (1972)
[4.31] J. Inoue, S. Maekawa: Phys. Rev. B **53**, R 11927 (1996)
[4.32] H.Y. Hwang, S.-W. Cheong, N.P. Ong, B. Batlogg: Phys. Rev. Lett. **77**, 2041 (1996)
[4.33] R.A. de Groot, F.M. Mueller, P.G. van Engen, K.H.J. Buschow: Phys. Rev. Lett. **50**, 2024 (1983)
[4.34] R.A. de Groot: Physica B **172**, 45 (1991)
[4.35] R.A. de Groot: Europhys. News **23**, 146 (1992)
[4.36] K. Schwarz: J. Phys. F **16**, L211 (1986)
[4.37] H. van Leuken, R.A. de Groot: Phys. Rev. B **51**, 7176 (1995)
[4.38] A. Yanase, K. Siratori: J. Phys. Soc. Jpn. **53**, 312 (1984)
[4.39] Z. Zhang, S. Satpathy: Phys. Rev. B **44**, 13319 (1991)
[4.40] A.A. Minakov, I.V. Shvets: Surf. Sci. Lett. **236**, L377 (1990)
[4.41] R. Wiesendanger, H.-J. Güntherodt, G. Güntherodt, R.J. Gambino, R. Ruf: Z. Phys. B **80**, 5 (1990)
[4.42] K.P. Kämper, W. Schmitt, G. Güntherodt, R.J. Gambino, R. Ruf: Phys. Rev. Lett. **59**, 2788 (1987)
[4.43] H. Brändle, D. Weller, J.C. Scott, J. Sticht, P.M. Oppeneer, G. Güntherodt: Int. J. Mod. Phys. **7**, 345 (1993)
[4.44] S. Blügel, D. Pescia, P.H. Dederichs: Phys. Rev. B **39**, 1392 (1989)
[4.45] R. Wiesendanger, D. Bürgler, G. Tarrach, A. Wadas, D. Brodbeck, H.-J. Güntherodt, G. Güntherodt, R.J. Gambino, R. Ruf: J. Vac. Sci. Technol. B **9**, 519 (1991)
[4.46] R. Wiesendanger, H.-J. Güntherodt, G. Güntherodt, R.J. Gambino, R. Ruf: Helv. Phys. Acta **63**, 778 (1990)
[4.47] R. Wiesendanger, D. Bürgler, G. Tarrach, H.-J. Güntherodt, G. Güntherodt: In *Scanned Probe Microscopy*, ed. H.K. Wickramasinghe, AIP Conf. Proc. (AIP, New York 1992) p. 241
[4.48] R. Wiesendanger, I.V. Shvets, D. Bürgler, G. Tarrach, H.-J. Güntherodt, J.M.D. Coey: Z. Phys. B **86**, 1 (1992)
[4.49] R. Wiesendanger, I.V. Shvets, D. Bürgler, G. Tarrach, H.-J. Githerodt, J.M.D. Coey: Europhys. Lett. **19**, 141 (1992)
[4.50] R. Wiesendanger, I.V. Shvets, D. Bürgler, G. Tarrach, H.-J. Güntherodt, J.M.D. Coey, S. Gräser: Science **255**, 583 (1992)
[4.51] R. Wiesendanger, D. Bürgler, G. Tarrach, T. Schaub, U. Hartmann, H.-J. Güntherodt, I.V. Shvets, J.M.D. Coey: App. Phys. A **53**, 349 (1991)
[4.52] R. Wiesendanger, D. Bürgler, G. Tarrach, I.V. Shvets, H.-J. Güntherodt: MRS Proc. **231**, 37 (1992)
[4.53] R. Wiesendanger, I.V. Shvets , D. Bürgler, G. Tarrach, H.-J. Güntherodt, J.M.D. Coey: Ultramicroscopy **42–44**, 338 (1992)
[4.54] I.V. Shvets, R. Wiesendanger, D. Bürgler, G. Tarrach, H.-J. Güntherodt, J.M.D. Coey: J. Appl. Phys. **71**, 5489 (1992)
[4.55] G. Tarrach, D. Bürgler, T. Schaub, R. Wiesendanger, H.-J. Güntherodt: Surf. Sci. **285**, 1 (1993)

[4.56] S. Iida, K. Mizushima, M. Mizoguchi, K. Kose, K. Kato, K. Yanai, N. Goto, S. Yumoto: J. Appl. Phys. **53**, 2164 (1982)

[4.57] S. Iida, M. Mizoguchi, N. lGoto, Y. Motomura: J. Magn. Magn. Mater. **31–34**, 771 (1983)

[4.58] E. Kita, Y. Tokuyama, A. Tasaki, K. Siratori: J. Magn. Magn. Mater. **31–34**, 787 (1983)

[4.59] J.M.D. Coey, I.V. Shvets, R. Wiesendanger, H.-J. Güntherodt: J. Appl. Phys. **73**, 6742 (1993)

[4.60] R. Wiesendanger, I.V. Shvets, J.M.D. Coey: J. Vac. Sci. Technol. B **12**, 2118 (1994)

[4.61] S.N. Molotkov: Surf. Sci. **261**, 7 (1992)

[4.62] R. Koltun: "Aufbau eines Ultrahochvakuum-Rastertunnelmikroskops und Untersuchungen des magnetischen Kontrasts auf atomarer und Mikrometer-Skala"; Ph. D. Thesis, RWTH Aachen (1995)

[4.63] R. Wiesendanger: In *New Concepts for Low-Dimensional Electronic Systems*, ed. by G. Bauer, F. Kuchar, and H. Heinrich, Springer Ser. Solid-State Sci., Vol. 111 (Springer, Berlin, Heidelberg 1992) p. 97

[4.64] R. Wiesendanger, M. Bode, M. Kleiber, M. Löhndorf, R. Pascal, A. Wadas, D. Weiss: J. Vac. Sci. Technol. B **15**, 1330 (1997)

[4.65] H.J. Elmers, J. Hauschild, H. Fritzsche, G. Liu, U. Gradmann, U. Köhler: Phys. Rev. Lett. **75**, 2031 (1995)

[4.66] M. Bode, R. Pascal, M. Dreyer, R. Wiesendanger: Phys. Rev. B **54**, R 8385 (1996)

[4.67] S.C. Hong, A.J. Freeman, C.L. Fu: Phys. Rev. B **38**, 12156 (1988)

[4.68] M. Pomerantz, J.C. Slonczewski, E. Spiller: Proc. Int. Symp. Phys. Magn. Mater. (World Scientific, Singapore 1987) p. 64

[4.69] M. Johnson, J. Clarke: J. Appl. Phys. **67**, 6141 (1990)

[4.70] S.N. Molotkov: JETP Lett. **55**, 173 (1992)

[4.71] R. Laiho, H.J. Reittu: Surf. Sci. **289**, 363 (1993)

[4.72] K. Sueoka, K. Mukasa, K. Hayakawa: Jpn. J. Appl. Phys. **32**, 2989 (1993)

[4.73] S.F. Alvarado, Ph. Renaud: Phys. Rev. Lett. **68**, 1387 (1992)

[4.74] S.F. Alvarado: Phys. Rev. Lett. **75**, 513 (1995)

[4.75] K. Mukasa: In *Proc. 1st Int. Symp. Adv. Phys. Fields*, ed. by K. Yoshihara (Nat. Res. Inst. Met., Tsukuba 1996) p. 125

[4.76] M.W.J. Prins, R. Jansen, R. van Kempen: Phys. Rev. B **53**, 8105 (1996)

[4.77] G. Nunes, Jr., N.M. Amer: Appl. Phys. Lett. **63**, 1851 (1993)

[4.78] Y. Manassen, R.J. Hamers, J.E. Demuth, A.J. Castellano, Jr.: Phys. Rev. Lett. **62**, 2531 (1989)

[4.79] Y. Manassen, E. Ter-Oranesyan, D. Shachal, S. Richter: Phys. Rev. B **48**, 4887 (1993)

[4.80] S.N. Molotkov: Surf. Sci. **264**, 235 (1992)

[4.81] S.N. Molotkov: Surf. Sci. **302**, 235 (1994)

[4.82] D. Shachal, Y. Manassen: Phys. Rev. B **44**, 11528 (1991)

[4.83] R. Prioli, J.S. Helman: Phys. Rev. B **52**, 7887 (1995)

[4.84] R. Wu, A.J. Freeman: Phys. Rev. Lett. **69**, 2867 (1992)

[4.85] H. Ness, F. Gautier: Phys. Rev. B **52**, 7352 (1995)

5. Photon Emission from the Scanning Tunneling Microscope

R. Berndt

With 28 Figures

Much of the fascination of Scanning Tunneling Microscopy (STM) results from its ability to directly image atomic and molecular structures of a wide variety of materials. Since its inception, one purpose of STM has been to go beyond imaging by performing local experiments with individual nanoscopic objects addressed by the STM tip [5.1]. The concept of the technique to be presented in this chapter is to use the tip of a STM as a source of low-energy electrons or holes to locally excite photon emission and thus study luminescence[1] phenomena on nanometer-sized structures of metals, semiconductors and molecules [5.2]. This combination of STM with optical methods offers several attractive features. First, photons represent a particularly versatile channel of information in addition to the tunneling current. Their intensity, spectral distribution, angular emission pattern, polarization status, and time correlation are accessible by sensitive optical methods and represent unique probes of the tunneling region. Spatial mapping of these physical quantities permits the addition of "true color" to STM images. Second, the emission of visible or infrared light is a chemically specific characteristic of many excitations of solids and molecules. This encompasses phenomena such as plasmons and interband transitions in metals, intrinsic luminescence and luminescent defects in semiconductors, molecular fluorescence, and modified optical properties owing to quantum-size effects in metal and semiconductor particles and nanostructures. The goal to study the luminescent excitations with high spatial resolution and to make use of them for identification of, e.g., individual molecules or luminescent defects justifies substantial effort. Third, photon emission is a signature of inelastic processes such as inelastic tunneling (IET) and hot-electron thermalization (Fig. 5.1). These processes, which are of fundamental and practical interest [5.3, 4], are extremely difficult to detect in STM by conventional tunneling spectroscopy due to their small contribution to the total tunneling-current. Photon emission provides a purely inelastic signal which is not veiled by noise of the elastic tunneling channel. Finally, the tunneling-current filament in the STM is a unique and well-defined excitation source for such experiments. Its lateral extent is of atomic dimensions and its distance from the sample may be controlled with picometer precision. The control of this distance is a crucial factor in

---

[1] We will use the words "luminescence" and "photon emission" interchangeably throughout this review.

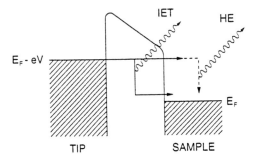

$E_F - eV$

IET   HE

E_F

TIP   SAMPLE

Fig. 5.1. Schematic energy diagram of a tunneling junction between two metal electrodes (E_F: Fermi energy of the right electrode, V_t: bias voltage of the left electrode, e: electron charge). Two possible mechanisms for photon emission are depicted: *solid line*: InElastic Tunneling (IET), *dashed line*: Hot-Electron Thermalization (HE)

any technique which probes the electromagnetic near-field of a sample. Using tunneling for distance control, the lateral resolution achieved in STM induced photon emission is superior to any other technique that involves visible light. To date, resolution of individual molecules in monolayers [5.5], atomic rows on metals [5.6] and nanometer sized quantum-well structures on semiconductors [5.7] have been demonstrated.

Photon emission stimulated by electron tunneling was discovered in 1976 by *Lambe* and *McCarthy* [5.8] from metal-oxide–metal junctions. Theoretically the emission was attributed to radiative decay of plasmons. These plasmons were excited either by inelastic tunneling [5.9, 10] or by a hot-electron mechanism in which an electron tunnels elastically and then thermalizes inside the collector electrode via plasmon generation [5.11]. The question as to which of these two possible processes dominates still appears not to be entirely solved. In 1972 *Young* [5.12, 13] had proposed the detection of light and secondary electrons generated by field emitted electrons in the topographiner, a precursor of the STM. The first experimental evidence of STM induced light emission from metals and semiconductors was published in 1988 by *Gimzewski* et al. [5.2] who detected ultraviolet photons from Ta and Si(111) surfaces. Soon afterwards, enhanced emission of visible light from Ag films was reported by *Gimzewski* et al. [5.14, 15].

Following this introduction, some practical aspects of STM induced photon emission are addressed in Sect. 5.1.

In any experiment that uses a macroscopic probe to investigate a single nanometer sized object probe–sample interactions are an important issue. As will be discussed in Sect. 5.2, the high photon intensities observed from metal surfaces are in fact due to the electromagnetic interaction of tip and sample. Photon emission therefore has been used to directly measure this interaction and its variation on an Ångstrøm scale. The various optical and tunneling spectroscopic techniques available have permitted the identification of many other factors that affect the photon emission such as the local dielectric function of the sample, the density of initial and final states for inelastic tunneling processes and electron-interference phenomena in the tunneling gap. Within a framework established by spectroscopic results and theoretical modelling, spatial maps of the photon emission intensity will be analyzed.

CathodoLuminescence (CL) stimulated with a Scanning Electron Microscope (SEM) is widely used for the micro–characterization of semiconductor materials and devices. The lateral resolution obtainable in CL is often limited by the large thermalization volume of the energetic electrons used in SEM. In Sect. 5.3, luminescence from semiconductors induced by a STM is discussed. This method offers distinct advantages. Electron tunneling from or to the tip provides a bright and extremely localized source of electrons or – unique to STM – holes. Moreover, the energy of these electrons is only a few eV and thus permits nanometer resolution in photon emission mapping of semiconducting materials and heterostructures. Moreover, the spin of tunneling electrons can be analyzed from the polarization of the emitted light.

Section 5.4 is devoted to the new and interesting phenomena which occur when molecules are inserted between tip and sample. These phenomena encompass molecular excitations, the interaction of the molecular states with the metal surfaces, and the modifications of the molecular photon emission within a microscopic cavity. Experimentally, molecular resolution in STM induced photon emission has been achieved from C_{60} monolayers and also from anthracene.

5.1 Experimental Considerations

The nanometer sized region between tip and sample may be considered a point source that is emitting photons into the hemisphere above the sample. In many conventional STM systems, the requirement for mechanical rigidity leaves little room for a photon-collection system. However, the low photon intensities often encountered make a large collected solid angle one of the most desirable features of such a system. To ensure maximum intensity at the detector, an optical system should compress the angular distribution to the limited acceptance angle of optical analysis equipment. Further aims are to achieve a wide working range of photon energies and a variety of spectroscopic and microscopic modes of operation. In addition, sufficient structural stability for atomic-resolution work must be maintained, mechanical access to the STM for tip or sample manipulation should not be compromised and the use of an optical microscope is highly desirable for observation of the tip-sample region.

5.1.1 Photon Collection

The direct way to maximize the detected solid angle is to place a photomultiplier (or an optical fiber) close to the tip region. This approach has mainly been employed in experiments performed under ambient conditions [5.16–18]. For operation in Ultra-High Vacuum (UHV), either lenses or ellipsoidal reflectors have been used to focus light from the tip region to a detector outside the vacuum system [5.7, 15, 19, 20]. With a low f-number condenser lens close to the STM inside the vacuum, a collection angle $\Omega \approx 1$ sterad has been achieved [5.19]. A more complex design is to surround the STM by an ellipsoidal reflector. Light emitted from the tip, which is located in one focal point of the ellipsoid, reaches the second focal point, which is placed outside the vacuum system. Thus, light emission into a solid angle $\Omega \approx 3$ sterad has been collected [5.19]. Whereas lens designs offer greater ease of use and superior focusing, which is important when a spectrometer is used; an ellipsoidal mirror collects photons from a substantially larger solid angle and is free of chromatic aberrations. Further details of STM design for use in UHV and the photon detectors employed are discussed in [5.19] and therefore are not included here [5.21].

Angular distributions of the photons emitted from metals are described fairly well by the emission pattern of an oscillating electric dipole above a surface. Most of the light is emitted at an angle of $\approx 30°$ with respect to the surface into an azimuthal ring of $\approx 30°$ polar width [5.22]. Circular polarizations of the emitted light have been measured using a Pockelscell and a polarization filter [5.23].

So far it has been assumed that light is detected on the tip side above the sample. In some experiments on thin metal films a different approach has been used; a weaker photon signal can also be measured from the backside of these films. To this end, films have been evaporated directly onto prisms [5.24] or onto optical fibers [5.25].

5.1.2 Modes of Measurement

In order to isolate the factors that affect the photon emission, one can take advantage of a variety of modes of measurement. These modes are described briefly below and the terminology used throughout this review is defined. *Photon maps* are obtained in a constant-current mode simultaneously with topographic data by recording the photon intensity for each pixel of the image using photomulipliers or avalanche diodes. Typical experimental parameters are tip voltages $|V_t| \gtrsim 1.5\,\text{V}$, tunneling currents $I_t \gtrsim 1\,\text{nA}$, and counting times of $\approx 10\,\text{ms}$ per pixel. Wavelength-resolved *luminescence spectra* contain direct information on the light-emitting transition. In this case detection involves a grating spectrometer and a multichannel detector which cause a lower overall sensitivity than that of a photomultiplier. Consequently, elevated integration times or tunneling currents are required for adequate counting statistics. By limiting the spectral response of a photomultiplier or avalanche photodiode using optical bandpass filters, *isochromat spectra* are measured as a function of tip voltage V_t at constant tunneling current I_t to study the influence of the excitation energy (eV_t) on the intensity of a particular emission feature. During the acquisition of isochromat spectra the tip-sample distance s is varied by the STM feedback loop to maintain I_t constant. In certain cases (Sect. 5.2.3.2) it is desirable to obtain additional information by separating the effects of s and V_t. This can be achieved by simultaneously measuring the photon intensity and conventional current–distance (I–s) or current-voltage (I–V) tunneling characteristics.

5.1.3 Photon Intensities

In practice, for clean Ag films in UHV count rates of $\approx 10^4$ counts per second (cps) at nA tunneling currents have been recorded using detection solid angles $\Omega \approx 0.2$ sterad [5.15, 26]. This is in reasonable agreement with theoretical estimates of the expected quantum efficiency of $\eta \approx 10^{-4}$ photons/electron [5.27, 28]. For experiments under ambient conditions, significantly lower values have been reported [5.16–18]. Still, the count rates have been suitable for photon mapping of the light detected as a function of lateral position of the tip on the surface. Concerning the cathodoluminescence of direct bandgap semiconductors, the estimated yield for the GaAs(110) surface is $\eta \approx 10^{-4}$ photons/electron in the tunneling mode [5.7]. For the CdS(11$\bar{2}$0) surface values of $\eta \approx 10^{-5}$ have been reported [5.29]. In both of the above examples photon mapping and optical spectroscopy have been achieved.

5.2 Photon Emission from Metals

5.2.1 Introduction

An energetic electron impinging on a metal surface can produce electromagnetic radiation via mechanisms such as transition radiation, surface plasmon excitation and bremsstrahlung. As demonstrated by *Berndt* et al. [5.30] this radiation can be stimulated by a STM tip operated in the *far field emission regime* ($|V_t| \gtrsim 100\,\mathrm{V}$) (Fig. 5.2). In experiments using conventional electron

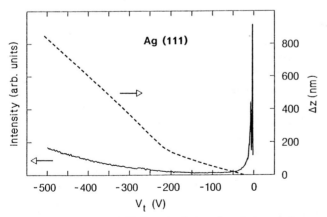

Fig. 5.2. Photon yield from a Ag surface integrated over the wavelength range 350–750 nm as a function of tip bias voltage V_t. The STM was operated at constant current $I_t = 1\,\mathrm{nA}$. The *dashed line* shows the simultaneously recorded tip excursion [5.30]. © American Physical Society 1991

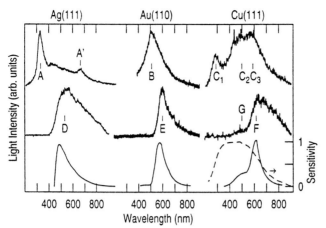

Fig. 5.3. Emission spectra from Ag(111), Au(110) and Cu(111). Spectra in the *topmost row* were observed in the high-voltage field-emission regime. Spectra in the tunnel regime ($V_t = 2.8$, 3.0, 3.6 V, $I_t = 10$, 10, 100 nA) are shown in the *middle row*. The results of a theoretical calculation for the emission in the tunneling regime using experimental parameters and a tip radius of 30 nm are presented in the *bottom row*. The sensitivity of the detection system shown as a *dashed line* was included in the calculation [5.30]. © American Physical Society 1991

guns the photon emissions ceases at low electron energies. In the STM, however, a new phenomenon has been observed [5.15]. Intense emission with quantum efficiencies as high as 10^{-4} photons/electron occurs at low bias voltages ($|V_t| \lesssim 40$ V), which corresponds to the *regimes of proximity field emission and tunneling* [5.15, 30]. This surprising observation called for further investigation, in particular because the highest intensities have been found in the voltage regime that is most useful in STM. The two distinct regimes of photon emission have been further characterized by luminescence spectra. The features in spectra from noble metals measured in the far field emission regime (Fig. 5.3, top row) correspond to surface plasmons (A) and various interband transitions (B, C_1, C_2, C_3). Strikingly different spectra observed in the tunneling and proximity field emission modes (see middle row of Fig. 5.3) consistently exhibit a shift of the main emission peak to longer wavelengths [5.14, 30]. The measured curve of tip-sample distance in Fig. 5.2 (dashed line) indicates the physical reason underlying the different photon emission characteristics. Whereas tip and sample are well separated at large $|V_t|$ the wavelength of the emitted light is much longer than the gap spacing in the tunneling regime. Therefore, the resulting electromagnetic coupling of tip and sample is expected to cause new emission characteristics.

Figure 5.4 displays luminescence spectra reported for Ag films and W tips for tip voltages in the tunneling and proximity field-emission regimes. The spectra show a dominant maximum of detected emission. At small bias voltages V_t the highest photon energy E_c observed is limited by the electron energy to $E_c = eV_t$, as expected. For $V_t \gtrsim 3$ V, the position of the major peak

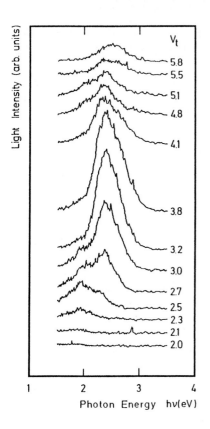

Fig. 5.4. Luminescence spectra measured from Ag films using W tips. The tip voltage for each spectrum is indicated. No correction for detector response has been made. © Les Editions de Physique 1989

is almost invariant with respect to the excitation voltage. This observation and the resemblance of the observed features with those observed from solid state tunneling junctions led *Gimzewski* at al. [5.14] to assign the emission to a resonantly enhanced process involving local plasmons. For a conventional inverse photoemission-related process, as observed for UV photons [5.2], the peak position would directly depend on the energy of the incident electron, in contradiction to the experimental observations.

5.2.2 Comparison with Theoretical Models

The experimental geometry in STM consists of the Ångstrøm sized gap separating two macroscopic bodies. A theoretical description of STM induced photon emission on metal surfaces has to treat the electromagnetic modes of this system. Moreover, the tunneling current and its interaction with these modes have to be dealt with. In the context of tunneling induced light emission from solid state tunneling junctions, two alternative excitation mechanisms had been invoked: first, inelastic tunneling occurring in the tunneling gap [5.8] and second, elastic injection of hot electrons and subsequent inelastic scattering within the sample [5.11]. The theories for STM-induced photon emission developed to date have approached these problems using similar models which approximate the complex experimental situation and provide a clear picture of the essential physics.

Persson and *Baratoff* [5.28, 31] considered tunneling from an *s*-like orbital at the apex of the tip to a spherical free-electron-like metallic particle. This model geometry permits the probabilities of competing radiative and non-radiative processes to be estimated analytically. While the yields from direct processes for photon emission (e.g., transition radiation) have been found to be low, enhanced emission occurs if the tunneling electron excites the dipole plasmon resonance of the metal particle. Processes involving dipolar plasmon excitation are summarized in Fig. 5.5. The large magnitude of the dipole moment associated with the plasmon results in a relatively high probability ($\approx 10^{-2}$) of inelastic tunneling processes (Fig. 5.5a). While dipolar plasmon excitation via inelastic tunneling is relatively improbable compared to elastic tunneling (≈ 1), it is clearly predominant compared to excitation via hot–electron decay ($\approx 10^{-4}$) (Fig. 5.5b). Once excited, dipole resonances can decay into electron-hole pairs (≈ 1) or via photon emission ($\approx 10^{-1}$) with a net yield given by the product of the relevant branching ratios. This results in overall estimated probabilities of only $P \approx 10^{-5}$ for hot–electron decay, but $P \approx 10^{-3}$ for inelastic tunneling. This probability for photon emission via inelastic tunneling is in reasonable agreement with experimental data for granular Ag films ($P \approx 10^{-4}$ photons/electron) [5.14, 26, 30]. The emission

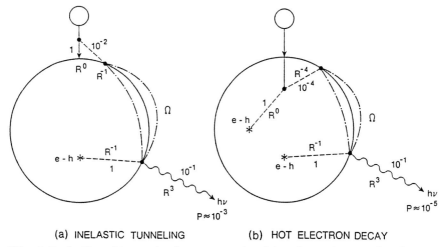

(a) INELASTIC TUNNELING   (b) HOT ELECTRON DECAY

Fig. 5.5a, b. Branching ratios for processes involving the excitation of a plasmon mode of frequency Ω (schematically indicated by *dash-dotted lines*) on a spherical metallic particle of radius R. The R-dependencies (R^n), typical estimates of branching probabilities for $R \approx 20\,\mathrm{nm}$ and the final photon yield P are indicated for (a) inelastic tunneling and (b) elastic tunneling followed by hot–electron decay. Competing non-radiative channels involving electron-hole pair creation are denoted by *e-h* [5.26]. © Kluwer Academic Publishers 1990

estimated for hot–electron decay occurring in the sample as invoked in the context of light emission from metal-oxide–metal tunneling junctions [5.11] is significantly lower and cannot explain the experimental findings in the STM experiments [5.28].

A second model developed by *Johansson* et al. [5.27, 32–34] takes the role of the tip into account. As in the first theoretical study of light emission from metal-insulator-metal tunneling junctions [5.9, 10] tunneling from a metallic sphere with radius R to a planar metal surface has been considered. The classical electromagnetic response (including plasmon resonances) of the model system has been calculated to obtain the strength of field fluctuations responsible for spontaneous emission. In the model, a quantum efficiency $P \approx 10^{-4}$ photons/electron is obtained for a W tip on a Ag surface in accordance with experimental results. Moreover, calculated luminescence spectra reproduce many features of experimental spectra and will be discussed below. According to this model, the reason for the enhanced emission is the formation of coupled, localized plasmons of the tip and the sample. These Tip-Induced Plasmon (TIP) modes are analogous to the symmetric or antisymmetric charge density oscillations observed on two metal-dielectric interfaces which are brought into close proximity [5.35]. The lowest-order mode, which is approximately dipolar, gives rise to a high probability of inelastic tunneling and photon emission (Fig. 5.6). Both models lead to a similar physical picture of STM induced photon emission. The experimentally observed intensities are consistent with inelastic tunneling excitation and radiative decay of a localized dipolar plasmon. In the model of [5.27] this plasmon is shown to result from the electromagnetic coupling of tip and sample (TIP modes).

Berndt et al. [5.30] obtained further evidence for this interpretation from a comparison of experimental luminescence spectra and model calculations (Fig. 5.3). The calculated spectra (bottom row of Fig. 5.3) reproduce essential features of the experimental data. Agreement between experiment and theory has been found in the positions of the maxima, the cutoff wavelength and the signal intensities. At long wavelengths, deviations between the model calculation and the experimental spectra are observed which are most likely due to radiation damping and retardation effects.

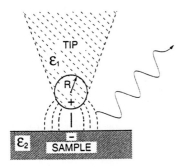

Fig. 5.6. Schematic model of a tip-induced plasmon mode in the STM configuration. In the model of [5.27] the tip is replaced by a sphere with radius R (typically 30 nm). Tip and sample are characterized by their complex dielectric functions ϵ. The + and − symbols indicate the approximately dipolar field of the lowest TIP mode

In the model of [5.27] most of the structure in calculated spectra results from the field enhancement G which describes the electromagnetic response of the tip-sample region. The structure of G is dominated by the complex dielectric function ϵ of the material with the smaller imaginary part of ϵ (lower damping), i.e., the noble metal surface when a tungsten tip is used. The role of the tungsten tip is to break lateral symmetry, thus giving rise to the tip-induced mode.

Figure 5.7 provides an overview of luminescence spectra obtained from noble metal surfaces using W tips for tip voltages covering the tunneling and proximity field emission regimes. All spectra show a single maximum of detected emission which exhibits a sharp cutoff at short wavelength and a broad tail in the red region of the optical spectrum. On Cu, a shoulder (S) is observed on the short wavelength side of the emission maximum. At small bias voltages V_t the shortest photon wavelength λ_c observed is determined by the energy of the tunneling electrons. The variation of the peak intensities with V_t in Fig. 5.7 will be discussed in Sect. 5.2.3. For Cu and Au surfaces the degree of agreement found between experiment and calculation is satisfactory for the entire range of bias voltages covered in Fig. 5.7 [5.33]. Correspondence, however, is worse at $|V_t| \gtrsim 4\,\mathrm{V}$ for Ag since a second maximum in intensity and shorter cutoff wavelengths are predicted by the model [5.27]. No indication of this second mode has been observed experimentally (Fig. 5.7) [5.22, 30]. In particular, no emission was detected on Ag from bulk and surface plasmons ($\lambda \approx 330\,\mathrm{nm}$). A possible explanation for this shortcoming of the model in-

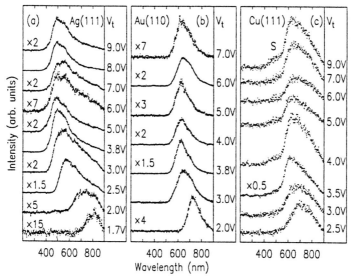

Fig. 5.7. Luminescence spectra obtained from **(a)** Ag(111), **(b)** Au(110), and **(c)** Cu(111) surfaces using W tips. The tip voltage for each spectrum is indicated. The spectra have been normalized to similar height for each metal separately using the factors noted in the spectra. No correction for detector response has been used

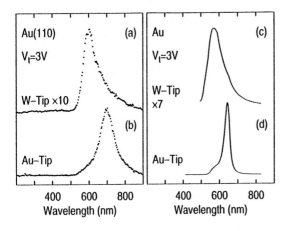

Fig. 5.8. Effect of tip material on luminescence spectra from Au(110) surfaces. (*a*) and (*b*) display experimental results for W and Au tips. (*c*) and (*d*) show the theoretical findings for identical bias voltages. The spectra have been normalized using the factors given in the figure [5.36]. © American Physical Society 1993

volves the non-spherical shape of a real tip and absorption of light by the Ag surface which drastically increases in the range of interest [5.30].

The important role in photon emission played by the tip manifests itself in modifications of emission spectra depending on the tip material as shown in Fig. 5.8 for Au(110) surfaces [5.36]. Spectrum (a) has been measured with a W tip and exhibits a single maximum. For a Au-covered tip (Fig. 5.8b) strikingly different emission characteristics have been found. The observed red-shift, the change in spectral shape, and the increase in emission intensity are reproduced in the theoretical model (Fig. 5.8c, d). These observations can be qualitatively understood as follows [5.36]. The W tip has no well-defined plasmon modes of its own and its role in the formation of the interface plasmon is passive. Au tips, on the other hand, have unique plasmon resonances, which on approaching a Au sample, couple with sample plasmon modes. Hence, coupled TIP modes acquire lower frequencies. The more intense emission from Au tips is mainly due to the smaller internal damping in Au. This interpretation is corroborated by similar findings for W and Ag tips on Ag(111) [5.36, 37]. A more intense, red-shifted and sharper emission peak is found with Ag tips. Moreover, additional spectral peaks at shorter wavelengths occur which were attributed to higher-order coupled modes having nodes inside the cavity. Interestingly, spectra measured with Ag-covered tips on Ag(111) [5.36] resemble early measurements with W tips on Ag films (Fig. 5.4) [5.14, 26] which may indicate that tips used in those experiments had unintentionally been covered with silver.

Light emission from transition metals was first investigated in [5.38]. While the large imaginary parts of the dielectric functions of these materials cause strong broadening of the TIP resonance, the observed luminescence and isochromat spectra are consistent with the model of TIP modes and IET excitation.

Recently, the model of [5.27] has been extended to magnetic substrates in order to interpret a circularly polarized component in the emission from ferromagnetic films [5.39]. While quantum effects such as spin polarized tunneling were neglected, the off-diagonal elements of the dielectric tensor which are responsible for the mixing of s and p polarization were taken into account. A reasonable agreement between experimental and calculated degrees of circular polarization was achieved suggesting that local magnetic properties may be analyzed using STM-induced photon emission.

In the models of [5.27, 28], the high frequency spectrum of the current $I(\omega)$, which excites the plasmon resonances, has been approximated by $I(\omega) \sim$ eV $- \hbar\omega$ [5.9, 10]. *Tsukada* et al. [5.40] and *Shimizu* et al. [5.41] have scrutinized the role of the spectrum $I(\omega)$ and studied its influence on the emission spectra and photon maps. In their first-principles calculation of the tunneling current spectrum small deviations of $I(\omega)$ from the linear approximation and a variation on an atomic scale were found.

In the experiments of *Takeuchi* et al. [5.24], STM stimulated light emission from the *backside* of metal films on gratings has been observed. *Ushioda* et al. [5.20] and *Uehara* et al. [5.42] applied models developed in the context of light emission from metal-oxide-metal tunneling junctions [5.9, 10, 43] to this geometry. Their results confirm the interpretation of light emission on the tip side as being due to TIP modes. Moreover, they found that some coupling of these modes to propagating surface plasmon polaritons occurs. The polaritons in turn emit light on the backside at specific angles determined by wavevector matching.

A possible role of the Tcherenkov effect in STM-induced photon emission has been put forward by *Smolyaninov* [5.44, 45] based on measurements of total photon intensity as a function of tunneling voltage conducted in air [5.46, 47]. Measurements in air are likely to be drastically affected by contamination and the resulting altered tunneling characteristics as well as by electrical breakdown and surface modifications at elevated voltages [5.17, 18]. We note that the model of *Johansson* et al. [5.27] does not predict a linear relationship between a normalized light intensity and $1/d$ (where d is the tip–sample separation) over an extended range of separations as assumed in [5.45].

5.2.3 Isochromat Spectroscopy

5.2.3.1 Proximity Field Emission Regime – Inelastic Tunneling Versus Hot-Electron Scattering. The excitation mechanism of tip-induced modes is particularly interesting since it involves the interaction of electron tunneling events with local electromagnetic modes. The model calculations [5.27, 28, 30] clearly favor inelastic tunneling which occurs in the tunneling gap as the excitation mechanism. This is in contrast with the situation for semiconductors where hot–electron decay within the semiconductor was found to dominate (Sect. 5.3). An independent method for demonstrating that inelastic processes are occurring in the tip–sample gap makes use of electron standing waves in the gap. These standing waves which were

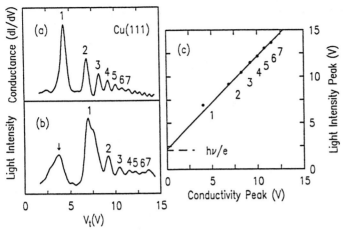

Fig. 5.9. (a) Differential conductance and (b) isochromat spectrum for photons with $\lambda = 600\,\text{nm}$ (b) measured simultaneously at constant current $I_t = 5\,\text{nA}$ on a Cu(111) surface using a W tip. The spectra have been obtained with a single voltage sweep in order to avoid smearing of the resonance structures. For peaks 1–7, which are related to field emission resonances, positions of maxima in photon intensity are plotted versus positions of maxima in differential conductance in (**c**). The experimental data (peak 1 excluded) is fitted by a *straight line* with slope 1 and intercept 2.2 V. The peak marked with an *arrow* is discussed in Sect. 5.2.3.2c [5.53]. © Johann Ambrosius Barth 1993

first observed in STM by *Binnig* and *Rohrer* [5.48] cause prominent oscillations in differential conductance (dI/dV) tunneling characteristics (Fig. 5.9) [5.49–52]. Their origin is illustrated in Fig. 5.10a, ba. An electron wave transmitted from the tip (left) through the potential barrier is partially reflected at the potential step at the surface of the sample (right) giving rise to interference. A series of field emission resonance states results at energies eV_n. When the Fermi level of the tip matches such a state, increased conductivity is observed which corresponds to an increased final density of states for tunneling. Isochromat photon spectra (Fig. 5.9b) exhibit similar oscillatory behavior [5.15, 30, 53]. In Fig. 5.9, the maxima in isochromat intensity are directly related to the conductance maxima by a constant offset of $eV_t = 2.2\,\text{eV}$ which is close to the energy of the detected photons of $\hbar\omega = 2.07\,\text{eV}$. Virtually identical results have been obtained with various combinations of tip and sample materials. Since the field emission states are confined to the gap region, *Berndt* and *Gimzewski* [5.53] concluded that the interaction of the electrons with tip-induced modes occurs within the tunneling gap.[2] For a hot–electron excitation one would expect electron injection and ballistic transport processes to be relatively insensitive to the matching conditions of the tip and surface wave functions. A related argument in favor of IET is

---

[2] The resonance states may also affect the time an electron spends in the gap region.

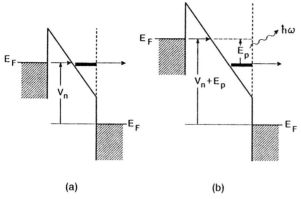

Fig. 5.10a, b. Simplified energy diagrams of a tunneling junction for two metal electrodes. The Fermi level of the electrode on the right is set to zero. In the field emission regime resonance states exist in the gap between the electrodes. (**a**) For applied voltages V_n that match the Fermi energy E_F of the negatively biased electrode with a resonance state, the differential conductance reaches maxima. (**b**) If the applied voltage is increased to $V_n + E_p/e$, E_p being the energy of a tip-induced mode, a tunneling electron can couple to the resonance state at eV_n by emitting a photon with an energy of $\hbar\omega = E_p$, which produces a maximum in photon intensity. © Johann Ambrosius Barth 1993

the absence of a significant bias polarity dependency of the emission spectra [5.26]. Moreover, isochromat spectra observed on semiconductor surfaces display distinctly different behavior (Sect. 5.3.2.2), which further supports the above interpretation.

5.2.3.2 The Tunneling Regime

a) *Electromagnetic interaction of two objects in nanometer proximity:* Optical emission from TIP modes provides a direct probe of the electromagnetic interaction of two bodies (the tip and the sample) in nanometer proximity. Surface-enhanced Raman scattering [5.54] is an example of a phenomenon that is critically influenced by similar local electromagnetic interactions. *Berndt* et al. [5.36] explored the field enhancement of TIP modes in the STM by varying the tip-sample distance in a controlled manner. Figure 5.11 displays the results of simultaneous measurements of I_t and the isochromat photon intensity as a function of tip-sample distance. Similar exponential dependencies of the tunneling current and the detected isochromat intensity on the distance were found. The measured photon intensity, however, was not exactly proportional to the tunneling current as emphasized in a plot of the photon yield (Fig. 5.11b). Despite the scatter of the data a decrease in yield with increasing distance is visible over the limited range of tip excursion used (0.2 nm). In the model of TIP modes, the intensity reduction per Ångstrøm calculated from the experimental parameters used (solid line in Fig. 5.11b) is $\approx 7\,\%$, which is in close agreement with the experimental observation [5.36].

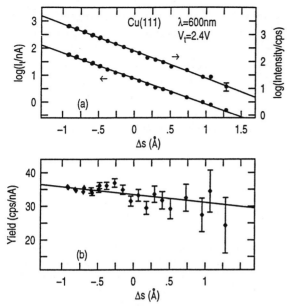

Fig. 5.11. (a) Simultaneous measurement of current I_t (*lower curve*) and isochro-mat photon intensity ($\lambda = 600$ nm) (*upper curve*) as a function of vertical tip displacement Δs at a fixed tip bias of $V_t = 2.4$ V. A W tip was used on a Cu(111) surface. The initial tip-sample distance ($\Delta s = 0$) is determined by the tunneling parameters set before disabling the STM feedback loop ($V_t = 2.4$ V, $I_t = 8$ nA). $\Delta s > 0$ corresponds to an increased tip-sample distance. *Straight lines* have been fitted to both data sets. (b) Yield (defined as intensity divided by current). The cal-culated result within the model of [5.27] (scaled by an arbitrary factor in y-direction and shifted in x-direction) is represented by a *solid line* [5.36]. © American Physical Society 1993

This small value results from two partially compensating effects. At larger tip-sample distances, the field enhancement is predicted to decrease while at the same time the tunneling electrons can interact with the TIP modes over longer distances. An exact solution of the Maxwell equations for a tip-surface geometry has recently confirmed the strong field enhancement associated with TIP modes and its sensitive dependency on the tip-sample separation [5.55]. As will be discussed in Sect. 5.2.4d, the distance dependence of the TIP modes leads to the atomic resolution observed in the photon emission from Au(110) surfaces.

b) *Barrier height for inelastic tunneling:* Conventional tunneling spectros-copy bears information on the elastic tunneling channels such as the tunneling barrier height or the density of states. In a similar manner, isochromat photon spectra may be used to study inelastic tunneling paths. The data in Fig. 5.11 has been used in [5.36] to directly compare the tunneling barriers of the elastic and inelastic channels. From the variation of the tunneling current

I_t with tip displacement Δs, an apparent tunneling barrier height Φ may be deduced [5.1]. The tunneling current is predominantly due to elastic tunneling whereas photons are generated by electrons which tunnel inelastically. This suggests that apparent barrier heights for elastic and inelastic tunneling, Φ_{ET} and Φ_{IET} can be derived from the measurements in Fig. 5.11. Surprisingly, these barrier heights are identical within the experimental uncertainty. This has led to the proposal that the inelastic processes occur preferentially in the region of the tunneling gap closest to the electron-collecting electrode [5.36]. A possibly related observation is known from inelastic tunneling spectroscopy of molecular vibrational modes, where a clear asymmetry in peak height in d^2I_t/d^2V_t spectra was observed depending on bias polarity [Ref. 5.4, p. 443].

c) *"Universal" intensity maximum at $|V_t| \approx 3.7\,V$:* In the tunneling regime an intensity maximum for $|V_t|$ between 3–4 V (arrow in Fig. 5.9b) has been observed on all metallic surfaces studied to date independent of the materials of both tip and sample and of the wavelengths of the detected photons [5.53]. Various explanation for this maximum had been proposed (e.g., nonlinear processes [5.15] or competition between localized and propagating plasmons [5.26]) none of which explains the universal occurrence of this maximum. As discussed by *Berndt* and *Gimzewski* [5.53], all observations are consistent with the following model which involves the competition of two counteracting mechanisms. First, in the tunneling regime, the initial increase in intensity can be attributed to the growing number of energetically allowed channels that can contribute to photon emission [5.26]. Second, the subsequent drop in intensity above $|V_t| \approx 3.5\,V$ has been suggested to result from a change in voltage-distance relation $s(V_t)$: a drastically stronger increase of $s(V_t)$ occurs for $|V_t| \gtrsim 3.5$–4 V [5.27]. This causes a smaller field enhancement below the tip, and thus a decrease in photon emission as the tip retracts in the constant-current mode with increasing $|V_t|$ [5.53]. This interpretation gained direct support from simultaneous measurements of I–V tunneling characteristics and isochromat spectra measured *at constant tip–sample distance* [5.53]. Under these conditions the photon intensity decrease in the range $|V_t| = 3$–4 V is absent. This demonstrates that the "universal" intensity peak is primarily due to the vertical tip motion occurring in that region.

The analysis above illustrates that structures in isochromat spectra at constant current result from the interplay of three effects. First, the tunneling characteristics of the elastic channel determine the tip-sample distance and hence the strength of the tip-induced modes which in turn affect the probability of inelastic tunneling. Second, the density of initial states that can contribute to the inelastic process plays an important role. Third, the density of states at the final-state energy, i.e., at $eV_t - \hbar\omega$, directly modifies the probability of inelastic tunneling.

Measurements of total photon intensity as a function of tunneling voltage conducted in air [5.16, 46, 47], have yielded results which differ from the above data obtained in UHV with isochromat spectroscopy. These differences are attributable mostly to the effects of contamination and the resulting altered tunneling characteristics as well as to electrical breakdown at elevated voltages. Under ambient conditions instabilities and surface modifications have been encountered on similar samples, which prevented reproducible measurements at elevated voltages [5.17, 18].

5.2.3.3 Polarization Effects. Recently, *Vasquez de Parga* and *Alvarado* [5.56] made the intriguing observation that a circularly polarized component is present in the photon emission from Co films and suggested that this polarization is related to magneto-optical effects. While the physical mechanism involved is not yet fully understood, the authors of [5.56] proposed that the photon spin and the spin polarization of the valence band of the ferromagnet couple via spin-orbit and exchange interactions. The observed dependence of the polarization on the azimuthal angle of emission indicates, however, that geometrical features of the tip apex play an important role as well [5.57].

A first theoretical description of this phenomenon has been presented by *Majlis* et al. [5.39]. By extending the model of TIP modes to magnetic materials they obtained a reasonable agreement between experimental and calculated degrees of circular polarization. In the calculation, quantum effects such as spin-polarized tunneling were neglected. *Majlis* et al. [5.39] concluded that measurements of the circular dichroism using STM may provide access to the optical constants of magnetic materials.

5.2.4 Photon Mapping

Photon maps are usually recorded in the constant current mode of operation where the *total* injected current is kept constant throughout the measurement. The prevailing contributions to this current are due to elastic tunneling channels; the inelastic contribution and the photon emission are therefore free to fluctuate. The IET-TIP model implies that a variety of electronic and geometric factors can affect the photon intensity on a local scale. Electronic properties determine the initial and final density of states available for inelastic processes and will thus modify the branching ratio between elastic and inelastic channels and hence the photon emission. Since the electromagnetic properties of the tip-sample region are determined by a field enhancement term which in turn modifies the probability of IET, geometric factors such as the tip-sample distance, the characteristic tip radius and the sample curvature determine the local TIP modes. The field enhancement also reflects the local dielectric function. The probabilities of non-radiative processes which compete with photon emission (such as electron-hole pair generation) may also vary locally and give rise to contrast.

a) *Granular films:* The first attempts of photon mapping were reported for thick silver films condensed onto Si(111) [5.15, 26]. Topographs of these films display a preferred (111) texture with grain sizes in the 20–50 nm range.

The corresponding photon maps clearly contained contrasting regions related to individual grains [5.26]. Some grains give rise to strong emission (exceeding 10 kHz/nA), while no emission was detected from other grains. Several groups have investigated STM induced emission from similarly structured polycrystalline films [5.16–18, 25, 47]. *Sivel* et al. [5.17] and *Gallagher* et al. [5.18] reported that the use of Au tips and samples was instrumental in achieving stable imaging and sufficiently high photon intensities under ambient conditions. In these studies a strong correlation of topographs and photon maps was observed for features on a scale of some 10 nm. An intriguing bias-dependency of the inter-grain contrasts on Ag films was reported by *McKinnon* et al. [5.58] who emphasized the importance of grain-boundary effects and of the interaction between grains. Recently, *Ito* et al. [5.59] have gone beyond photon mapping of granular films and characterized inter-grain variations of the emission from granular films spectroscopically. Inter-grain variations have also been observed in isochromat photon mapping of Au films [5.60]. Similar experiments on well-separated metal particles or clusters appear feasible.

The interpretation of images from granular films is complicated by several factors, most notably the structural complexity of the films and in particular by the columnar extension of the grains beneath the surface, which determines coupling between grains. Preferential adsorption of residual gases may contribute as well to intensity variations, such as distinct intra-island intensity variations observed on certain samples. Nevertheless, these images provide a feeling for the richness of factors that determine photon maps and differentiate them from topographs. Below, analyses of photon maps of well-defined surfaces, adsorption systems and local modifications introduced by the STM tip are presented. These samples represent nearly idealized systems and are thus more directly amenable to the theoretical models developed to date. In this way, various factors that determine the local efficiency of photon emission can be disentangled.

b) *Density-of-states effects:* Both the total tunneling current and photon intensity depend on the Density Of States (DOS) of tip and sample. The direct consequences for the interpretation of photon maps are as follows. For $V_t < 0$ V electrons tunnel from occupied levels of the tip to unoccupied states of the sample (Fig. 5.12). Due to the sharp decay of tunneling probability with barrier height, occupied states close to the Fermi level of the tip

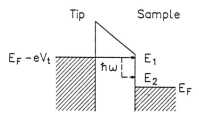

Fig. 5.12. Schematic energy diagram of a tunneling barrier for a negatively biased tip. *Solid line*: elastic tunneling; *dashed line*: inelastic tunneling (E_F: Fermi energy of the sample, V_t: tip voltage, $h\nu$: photon energy) [5.61]. © American Physical Society 1993

contribute most efficiently to I_t. Consequently, the unoccupied DOS of the sample at $E_1 = E_F - eV_t$ determines the elastic tunneling channel. Inelastic processes probe a different window of empty sample states centered around $E_2 = E_1 - h\nu$. Given the constant current mode of operation of the STM, variations in the electronic structure at E_2 govern the probability of inelastic tunneling.

Experimental evidence of the role of the local density of states in mediating contrasts in photon maps was reported for oxygen-exposed Ti films [5.38, 61]. In constant, current STM images (Fig. 5.13) the adsorption of oxygen on titanium gives rise to the formation of topographic structures that are dependent on bias voltage. These structures appear as protrusions at negative V_t and as shallow depressions at positive V_t. Steps show similar

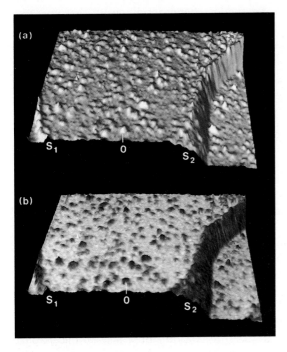

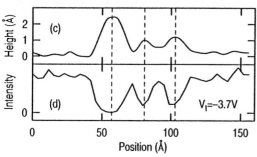

Fig. 5.13. (a) STM topograph of a Ti film exposed to 20 L of O_2 at room temperature (area: $100 \times 100\,\mathrm{nm}^2$, height: 2 nm, $I_t = 50\,\mathrm{nA}$, $V_t = -3.7\,\mathrm{V}$). Examples of steps (S_1, S_2) and oxygen-induced structures (O) are marked. (b) Photon map recorded simultaneously (intensity scale: 0–1 kHz). (c), (d) Cross sections through an area containing three oxygen-related structures. *Dashed lines* serve to guide the eye [5.38]. © Elsevier Science Publishers 1992

characteristics. In photon maps (Fig. 5.13b) recorded simultaneously with the topography (Fig. 5.13a) the emission intensity is significantly diminished at steps (S_1, S_2) and also at oxygen-related structures (see marker O for an example). It is notable that variations of photon intensity introduced by the oxygen-related structures on Ti are observed on a similar length scale to the topographic data. The high spatial resolution observed in these photon maps at oxygen-induced structures was proposed to arise predominantly from modifications of the unoccupied DOS of the sample [5.38, 61] although geometric effects (Sect. 5.2.4d) are likely to be involved as well. This interpretation is consistent with electron spectroscopic data of the filled and empty DOS [5.61].

The mechanism discussed above is likely to be present on the tip surface as well. As discussed in [5.61], the motion of tip atoms or adsorbates on the tip is probably responsible for sudden changes in intensity that give rise to stripes along the slow scan direction in the photon maps. These stripes were often found to be closely correlated with sudden subatomic changes in apparent surface height which vary randomly in repeated scans.

c) *Geometric effects: the* 10 nm *scale:* To study model systems such as small single particles on otherwise flat single-crystal surfaces *Berndt* and *Gimzewski* [5.62, 63] took advantage of the ability of the STM to modify surface structures locally. Figure 5.14 shows a protruding structure with apparent dimensions of 5 nm in height and some 25 nm in diameter generated on a Cu(111) surface. The real shape of a protrusion may differ from the STM topograph due to a convolution with the tip shape. While the detected photon intensity (Fig. 5.14b is fairly constant above flat Cu(111) regions, a pronounced increase in emission by approximately one order of magnitude is found when the tip is positioned over the protrusion. This local increase in intensity is in accordance with theoretical expectations of the intensity as a function of particle size. The model of *Persson* and *Baratoff* [5.28] predicted an increased photon emission from nanometer sized metallic particles relative to a flat surface. Furthermore, the sharp contrast occurring in the photon map is additional evidence that the emission is due to a localized mode between the tip and the protrusion. In the case of roughness-mediated emission from a propagating surface plasmon – the protrusion would act as a scattering center facilitating the conversion of plasmons into photons – a slow decay of intensity over the mean plasmon decay length on a micrometer scale would be expected. *Kroo* et al. [5.64] have demonstrated such a slow decay using a STM to detect propagating plasmon waves launched by photons in a Kretschman configuration. The increase in intensity at the protrusion provides some information on its chemical nature as well. It is consistent with a copper-containing protrusion. Tungsten from the tip that might have formed the structure when the voltage pulse was applied would significantly reduce the emission intensity due to its different dielectric function. The large imaginary part of the dielectric function of W is expected to lead to a broad and

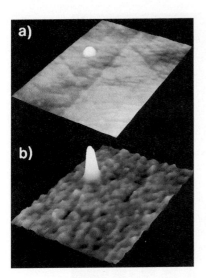

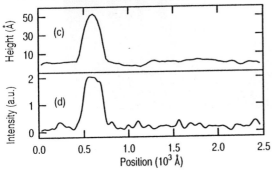

Fig. 5.14a–d. $160 \times 200 \, \text{nm}^2$ sub-area of a topograph of a Cu(111) surface ($V_t = 2.2 \, \text{V}$ and $I_t = 2 \, \text{nA}$). (a) Topograph (height scale: 5 nm) rendered as a pseudo-three-dimensional object. (b) Image (a) with photon intensity data superimposed as a grey scale coloration corresponding to 0–2000 Hz. (c), (d) Cross sections of original topograph and photon map, respectively [5.62]. © Elsevier Science Publishers 1992

relatively weak TIP mode as observed previously for other transition metals (Ti, Fe [5.38]).

d) *Atomic resolution in photon mapping:* The contrasts in photon maps associated with particles and surface structures in the 10 nm range may be understood in terms of the radius-dependent dipole moments and the dielectric functions of these structures. In addition, even finer topographic variations can significantly change the probability for photon emission [5.30, 61]. A well-defined test of the resolution capability of STM induced photon emission from metal surfaces was performed on a (1×2) reconstructed Au(110) surface [5.6]. This surface is comprised of atomic rows which served as an atomic-scale line grating. In a conventional STM image of a stepped surface area (Fig. 5.15a), terraces of close-packed gold rows separated by 0.81 nm and monoatomic steps are resolved. In the photon map recorded simultaneously at constant I_t (Fig. 5.15b), the same atomic periodicity is observed. Closer inspection reveals that intensity maxima correspond to topographic minima.

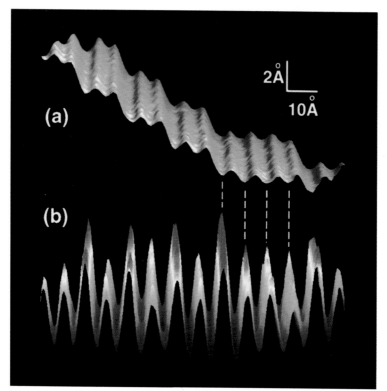

Fig. 5.15. Constant current STM image (**a**) and photon map (**b**) of a stepped area of a (1×2) reconstructed Au(110) surface measured simultaneously. In (**b**) the photon intensity is represented by height and grey level. The overall modulation of the photon intensity is ≈ ±25% of the average intensity. Close-packed gold rows along the [1$\bar{1}$0] direction which are separated by 0.81 nm are resolved in the topograph. An identical but phase-shifted periodicity is found in the photon map with increased photon intensity maxima near steps. Temperature of tip and sample: 50 K [5.6]. © American Physical Society 1995

The variation in photon intensity does not arise from spurious variations in I_t because no directly correlated structure is observable in simultaneously recorded maps of the tunneling current. Local variations of the density of states for inelastic tunneling or of the tunneling barrier can also be excluded as possible mechanisms mediating the atomic resolution, as was discussed in [5.6]. A description of all experimental data was obtained from an analysis of the electromagnetic field strength of the TIP modes which was motivated by the following two facts [5.6]. First, both tunneling current and photon intensity are known to depend sensitively on sub-Ångstrøm changes of the distance between tip and sample (Fig. 5.11) [5.36, 55]. Next, whereas the tunneling current is localized laterally to atomic dimensions, the TIP modes which mediate the photon emission have a lateral extension of several

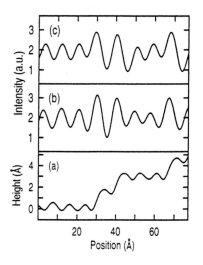

Fig. 5.16. Cross-sections through a stepped area of a Au surface and results of a model calculation. (**a**) Section of a STM topograph perpendicular to the Au rows. Several small terraces comprised of atomic rows and monoatomic steps are resolved. (**b**) Corresponding section of the photon map. The photon signal exhibits an atomic periodicity which is phase-shifted with respect to the topograph. On the lower side of steps, photon intensity maxima are significantly more intense. (**c**) Calculated photon intensity. Adapted from [5.6]

nanometers. As a consequence of this mismatch in lateral resolution, most surface structures that cause a vertical tip displacement will also modify the electromagnetic modes and hence the photon emission. This concept led to a phenomenological model where the photon intensity from a corrugated surface $h(x)$ is estimated by defining a surface of constant TIP mode strength from a convolution of the topographic surface with the lateral extent of the TIP mode [5.6]. For a small sub-Ångstrøm deviation of the tip from this surface of constant TIP mode, a linear intensity variation is expected (Fig. 5.11). The encouraging similarity of simulated emission patterns (Fig. 5.16) with the observed patterns suggested that despite its simplicity this model describes the essential physics underlying the observed atomic resolution. According to the above interpretation, the electromagnetic interaction of the tip and the sample as reflected by the photon emission from TIP modes varies on an atomic scale. The reason for these variations is the different lateral length scales relevant for tunneling and for the electromagnetic properties of the tip-sample region. This idea is valid in a more general context, as discussed by *Rohrer* [5.65]. Many scanning probe microscopies use two interactions of the tip and the sample. First, the proximity of sample and probe ("topography") is defined in terms of a *control interaction*. Second, a *measurement interaction* is monitored under the condition of constant control interaction. In the case of STM-induced photon emission these interactions are the tunneling current and the electromagnetic coupling. In order to separate the effects of these interactions, an accurate measurement and a detailed understanding of both interactions are prerequisites.

e) *New effects:* While the basic factors involved in the photon emission from metals have been identified and described, nevertheless there exist a number of effects that require further evaluation, in particular for complex systems such as amorphous or polycrystalline materials. Below, some of these phe-

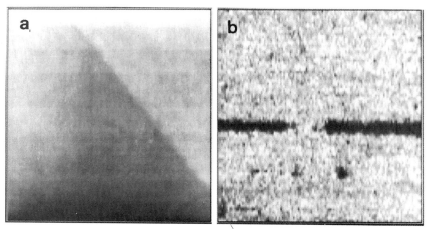

Fig. 5.17. Topograph (**a**) and photon map (**b**) from a facet of a Au ball ($I_t = 10\,$nA, $V_t = -1.75\,$V, area $= 120{\times}120$nm^2). A band of lower emission intensity was formed in a uniformly emitting area by scanning a line at $V_t = -2.10\,$V [5.66]. © American Physical Society 1994

nomena are discussed. The controllable and reversible manipulation of the photon emission characteristics of Au surfaces in air has recently been reported by *Sivel* et al. [5.66]. Figure 5.17 displays a photon map of an emissive area of a Au film where a 5 nm wide, non-emissive band has been generated at an elevated tip bias. The reverse process, writing of emissive features, was also demonstrated. *Sivel* et al. [5.66] discussed these experiments in terms of material transfer between the tip and the sample under the influence of the electric field. The role of adsorbate molecules has been treated in [5.67]. While the mechanism underlying writing and erasing has not yet been definitely identified, it appears likely that the phenomenon is specific to experiments under ambient conditions [5.66]. *Gimzewski* et al. [5.68] presented photon maps from thin Ag films on Si(111) exhibiting bright emission bands of nanometer-widths which did not correlate with the features of the surface topography.

5.3 Photon Emission from Semiconductors

As proposed in [5.2], the tip of a STM may be used as a bright and extremely localized source of electrons (or holes) to excite photon emission from semiconductors. The low energies of the electrons in a STM, which may even be below the vacuum level, permits nanometer lateral resolution to be achieved.

5.3.1 Indirect Gap Materials

Before presenting results for direct gap materials, which have been amenable to detailed interpretation, we briefly review experiments performed on semiconductors with indirect gaps. In their pioneering experiment, *Gimzewski* at al. [5.2] observed the emission of ultraviolet ($h\nu = 9.5\,\mathrm{eV}$) photons from a Si(111) surface. Isochromat spectra of the emission showed similarities with normal-incidence inverse photoemission spectra and additional features attributed to field-emission resonances or image states. *Ushioda* [5.69] reported spectra of visible luminescence from Si single crystal surfaces. The spectra and the polarization status depended on various factors including the sample doping, the tip material and the surface orientation. Yet no assignment of the electronic transitions involved was possible. One reason for this is that for the high electron energies used in this experiment ($E \approx 25\,\mathrm{eV}$), a complex cascade of small electronic transitions is energetically allowed. Photon mapping of porous silicon was undertaken by *Dumas* et al. [5.70, 71] and *Gu* et al. [5.72]. *Ito* et al. [5.73] recently reported a dependence of the spectra of STM induced light emission from porous silicon and interpreted this observation in terms of a quantum size effect.

Despite the high electron energies used in experiments on crystalline Si ($> 10\,\mathrm{eV}$ in [5.2] and $> 25\,\mathrm{eV}$ [5.69]) the quantum efficiencies for luminescence were low and large currents were required to obtain acceptable photon intensities. This problem is significantly less severe for direct gap materials. Sufficient photon emission can be excited by the STM at electron energies and currents which are less likely to induce surface damage.

5.3.2 Direct Gap Materials

5.3.2.1 Luminescence Spectra. STM stimulated photon emission of direct gap materials was first reported by *Abraham* et al. [5.7, 74] for p-type GaAs. Spectra of this emission in the photon energy range $h\nu = 1.4$–$3.5\,\mathrm{eV}$ exhibit a single peak at $h\nu \approx 1.45\,\mathrm{eV}$ which corresponds to electron-hole pair recombination between the GaAs Conduction Band (CB) and valence-band edges [5.75]. No contribution of the tip itself to the spectra was observed. On wide gap materials such as CdS, *Berndt* and *Gimzewski* [5.29, 75] found additional structure in luminescence spectra (Fig. 5.18). Peak G at 490 nm (2.53 eV) was attributed to luminescence generated via radiative recombination of electrons from levels close to the CB edge with holes in the vicinity of the valence-band edge. At room temperature the various contributions of radiation involving transitions between shallow impurities, exciton states and the electronic bands result in spectral broadening, which is consistent with the observed width of peak G. The broad emission band C observed around $\approx 700\,\mathrm{nm}$ (1.8 eV) indicates the presence of radiative deep impurities or defect levels. The relative strengths of these features were observed to vary with lateral tip position. Interestingly, the spectra show strong similarities with conventional cathodoluminescence at room temperature although the energy of the incident electrons is several orders of magnitude smaller in the STM experiment. Low-temperature luminescence experiments on GaP and AlGaAs–GaAs using STM have been reported by *Montelius* et al. [5.76, 77]

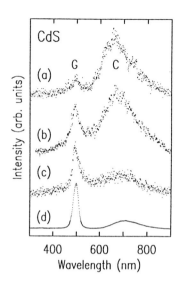

Fig. 5.18. Luminescence spectra observed at different tip positions on a CdS(11$\bar{2}$0) surface. The spectra have been acquired at $I_t = 10$ nA and $V_t = -7$ V in (a), (b), and (d) and $V_t = 10$ V in (c). A longer integration time was used for spectrum (d). The spectra are not corrected for instrumental response. G: intrinsic luminescence, C: extrinsic luminescence [5.29]. © American Physical Society 1992

and by *Samuelson* et al. [5.78, 79]. On AlGaAs–GaAs quantum wells, spectral changes were observed for different tip positions with respect to the quantum wells [5.79]. It should be noted that the high tunneling currents used are likely to cause surface damage or point contact formation [5.80, 81].

5.3.2.2 Isochromat Spectra. The luminescence spectra presented above demonstrated that the photon emission from semiconductors is due to interband recombination as well as deep impurities or defects. In contrast to the case of metals, no influence of the STM tip was identified. The excitation mechanisms involved in STM-induced photon emission from semiconductors have been revealed by isochromat spectra. Figure 5.19 displays the isochromat intensities ($h\nu \approx 1.5$ eV) acquired on p-doped (10^{18}–10^{19} cm^{-3}) GaAs when

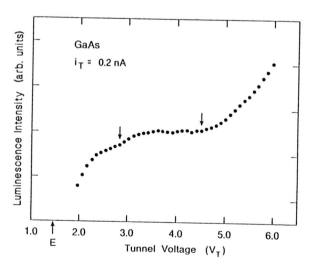

Fig. 5.19. STM-induced luminescence intensity vs sample bias on GaAs(110) [5.90]. © American Vacuum Society 1991

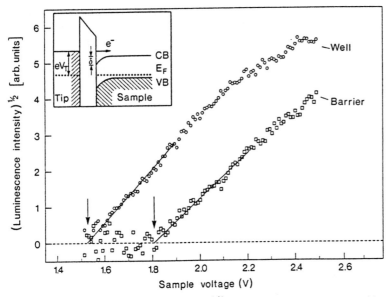

Fig. 5.20. (Luminescence intensity)$^{1/2}$ vs sample bias on an $Al_{0.1}Ga_{0.9}As/$ $Al_{0.3}Ga_{0.7}As$ heterostructure. The *upper data* was taken from a 100 nm well, and the *lower data* from a 50 nm barrier, both at a distance of 20 nm from the interface. The *inset* schematically shows an energy diagram of the tunneling barrier and the semiconductor surface. VB and CB are the edges of the valence and conduction bands, respectively, and E_F is the Fermi energy [5.87]. © American Physical Society 1991

injecting electrons into the semiconductor (for more detailed measurements of the threshold for emission cf. Fig. 5.20). For electron energies exceeding the GaAs bulk CB edge, emission is observed. This is consistent with tunneling of electrons from the tip into the GaAs CB and subsequent recombination with holes from the valence band. Additional weak features marked by arrows in Fig. 5.19 have been associated with the thresholds for the creation of electron-hole pairs. These thresholds can be recognized clearly in isochromat spectra of n-type CdS($11\bar{2}0$) (Fig. 5.21). For negative V_t (Fig. 5.21a) thresholds occur for $V_t \approx -4\,V$ and $V_t \approx -8\,V$. This observation was interpreted using an energy diagram of the tunneling junction shown in Fig. 5.22. For a negatively biased tip (a), the hot electron injected from the tip needs to create a hole by impact ionization to induce luminescence in the n-type material. Assuming parabolic bands and conservation of energy and momentum, the threshold energy for this process was estimated to be $E_{eh} \approx 3.9\,eV$ which is close to the experimental value. The threshold for creation of two pairs, 6.7 eV above the CB edge [5.82], is also comparable to the experimental threshold at 8 eV. The isochromat spectrum of Fig. 5.21a shows good qualitative agreement with conventional cathodoluminescence [5.83]. For positive V_t (Fig. 5.21b) a threshold for emission is found at $\approx 2.5\,eV$ close to the CdS

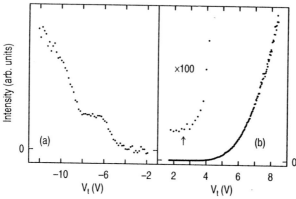

Fig. 5.21. Intensity of band-edge emission (isochromat spectra) as a function of applied tip bias, recorded at constant current $I_t = 1\,nA$ **(a)** Negative tip bias. **(b)** Positive tip bias. The low voltage region is enlarged and slightly offset in the vertical direction. As discussed in the text, a threshold of 2.4 eV (*arrow*) corresponding to the position of the valence band maximum is expected. Adapted from [5.29]. ©️ American Physical Society 1992

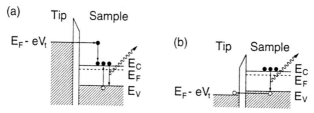

Fig. 5.22. Energy diagrams of the tunneling junction for **(a)** negatively and **(b)** positively biased tips. On materials with low conductivity strong tip-induced band bending is expected. It is schematically included in the figure. In **(b)** the effect of momentum conservation discussed in the text has been omitted for simplicity. (E_C: conduction-band edge, E_V: valence-band edge). Adapted from [5.29]. ©️ American Physical Society 1992

valence band edge. This is consistent with tunneling of electrons from the CdS valence band to the tip (Fig. 5.22b).

The photon intensity near the threshold has also been studied on GaAs surfaces coated with Au by *Wenderoth* et al. [5.84]. For p-type material threshold values close to the GaAs bandgap were obtained. On n-type material, the threshold values deviated from those expected for impact ionization. *Wenderoth* et al. [5.84] attributed this discrepancy to band-bending effects.

5.3.2.3 Photon Mapping and Applications

a) *Heterostructures: Abraham* et al. [5.7] investigated the STM-induced luminescence from $GaAs/Al_xGa_{1-x}As$ quantum-well structures. Using optical filters the detector response was restricted to GaAs luminescence ($E_G = 1.42\,eV$ for GaAs, $E_G = 1.9\,eV$ for $Al_{0.38}Ga_{0.62}As$). Thus, GaAs wells were resolved

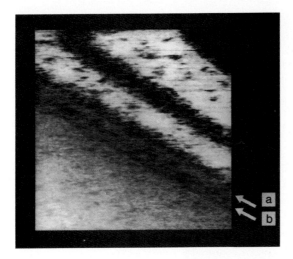

Fig. 5.23. Luminescence image of a multiple quantum well structure of GaAs/ $Al_{0.38}Ga_{0.62}As$. The central bright band is a GaAs well 10 nm thick. Quantum wells of 5 and 2 nm width are indicated by arrows a and b, respectively. Scan area: 78×78 nm^2 [5.7]. © American Institute of Physics 1990

as bright bands in the photon emission (Fig. 5.23) which were separated by regions of lesser intensity associated with tunneling into the AlGaAs barriers. An analysis of profiles of the photon intensity near wells revealed two characteristic decay lengths (≈ 10 nm and ≈ 400 nm). Whereas the latter decay length was energy independent, the former decreased with decreasing electron energy [5.85].

These observations can be rationalized in the following physical picture. Surface defects and contamination cause downward band bending on the p-type (Al)GaAs [5.86]. Moreover, they are expected to provide non radiative channels for recombination and thus quench luminescence from the surface region. Therefore, only electrons with energies above the bulk CB edge of GaAs reach the bulk GaAs where they can stimulate luminescence which will then be detected at $h\nu \approx 1.5$ eV. In this scenario, the emission is most intense when electrons are injected directly into the wells. On AlGaAs barriers, the probability for an electron to reach the bulk GaAs is expected to be governed by the thermalization length of hot electrons, which depends on the electron energy and the energy independent drift length of thermalized electrons. This explains the observed decay lengths and their variations with energy [5.85]. In the experiment of [5.85], highest contrast was observed for electron energies falling between the conduction band edges of GaAs and AlGaAs. This is consistent with the trapping of electrons in the surface band bending region of the AlGaAs under these conditions.

Within the framework of the above interpretation, the thresholds for exciting luminescence correspond to the bulk CB edges (Fig. 5.20). Spatial profiling of the threshold has been used by *Renaud* and *Alvarado* [5.87] to determine the position of the CB edge in multiple quantum well structures. A complication in interpreting these experiments were observed threshold values on GaAs which lie below the GaAs CB edge. These observations were tenta-

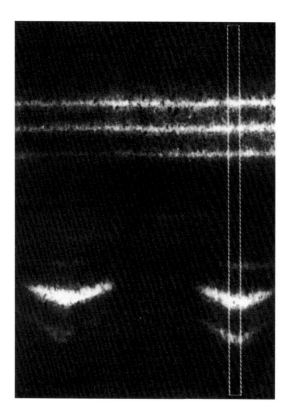

Fig. 5.24. Photon map of a $300 \times 300\,\mathrm{nm}^2$ cross section of a quantum well-wire array. The luminescence profile on the right shows the intensity in the boxed area, averaged horizontally [5.88]. © American Institute of Physics 1994

tively attributed to radiative recombination in the GaAs surface region under small band bending conditions [5.87]. Recently, *Pfister* at al. [5.88] extended luminescence mapping of heterostructures to quantum wires (Fig. 5.24). In luminescence spectra of these structures several quantum wells were found to contribute to the spectra simultaneously. This effect was attributed to the thermalization length of the injected electrons (30 nm), which is comparable to the separation of the quantum wells investigated [5.88].

b) *Surface effects:* So far, STM induced luminescence has been interpreted mainly in terms of the bulk band structure. The roles of surface and bulk effects were addressed in [5.29] for CdS surfaces. Figure 5.25 shows a topography and a photon map from a UHV cleaved CdS(11$\bar{2}$0) surface area that contains a sequence of steps (S). At the steps the luminescence intensity is drastically reduced. Possible mechanisms mediating this surface effect are a modification of the probabilities for electron-hole pair creation or the opening of nonradiative channels for energy dissipation. The sharpness of contrast in photon intensity suggests an electron injection related mechanism. An example of bulk contributions to the luminescence is the feature (C), a $\approx 4\,\mathrm{nm}$ diameter spot of higher emission intensity. This was observed on a surface region which is featureless down to a sub-Ångstrøm level and, therefore, was

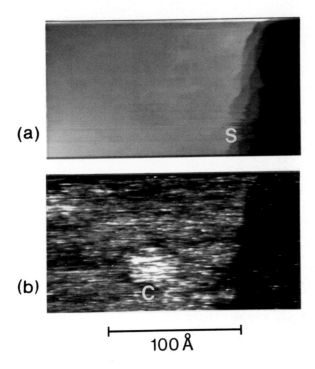

(a)

S

(b)

C

⊢————————————⊣
100 Å

Fig. 5.25. (a) STM topograph represented as gray-scale image taken from a $20\times10\,\mathrm{nm}^2$ region of CdS(11$\bar{2}$0) at $V_t = -7\,\mathrm{V}$ and $I_t = 5\,\mathrm{nA}$. (b) Photon map recorded simultaneously. Intensity scale: 0–1 kHz. [5.29]. © American Physical Society 1992

attributed to a subsurface defect. As shown in Fig. 5.18, defects give rise to intense features in the luminescence spectra. Zero-dimensional defects or dislocations are known to affect the image contrast in cathodoluminescence, as well [5.89].

The role of the STM tip in luminescence from semiconductors has so far been neglected. However, as in the case of metal surfaces, occasionally the emission intensity from semiconductors can change by an order of magnitude [5.90]. These variations may be due to structural or chemical changes of the tip which, in turn, change tip-induced band bending or the energy spectrum and angular spread of the tunneling current.

c) *Polarization effects:* GaAs(110) surfaces can be used as spin detectors because of the spin-orbit splitting of the valence band. In STM induced luminescence this gives rise to a particularly intriguing "tip effect". With Ni tips *Alvarado* and *Renaud* [5.23] detected the emission of circularly polarized light from GaAs surfaces which they interpreted in terms of spin polarized tunneling of 3d-electrons from the Ni Fermi level. The electron spin polarizations deduced from the circular polarization data were significantly higher than in conventional spin-polarized field emission. From the observed polarization increase with increasing tunneling current *Alvarado* [5.80] concluded that the relative contributions of strongly spin-polarized 3d electrons and low-polarized 4sp electrons vary with the width of the tunneling gap.

d) *Time-resolved measurements:* Photon mapping of GaAs surfaces under ambient conditions was investigated by *Bischoff* et al. [5.91, 92] and Horn [5.93]. Recently, these experiments have been extended to perform time-resolved measurements. To obtain a pulsed excitation of the tunneling induced luminescence voltage, pulses were added to the tunneling bias and the time between a pulse and the subsequent arrival of the first photon was measured. The observed decay times ($\approx 250\,\mathrm{ns}$) have been analyzed in terms of carrier lifetimes and dopant concentrations [5.94, 95].

5.4 Molecules on Surfaces

The discussion of the photon emission from metal surfaces in the STM made evident the important role played by the TIP modes of the cavity formed by the tip and the sample. New phenomena are observed when molecules are placed in this cavity [5.5, 96, 97]. In photon maps of close-packed C_{60} monolayers on Au(110) *Berndt* et al. [5.96, 97] resolved individual molecules as distinct maxima (Fig. 5.26). It should be noted that the intermolecular distance in these monolayers is $\approx 1\,\mathrm{nm}$. The striking observation of Fig. 5.26 indicates that the STM may be used to selectively address single molecules and measure their photon emission.

As discussed in Sect. 5.2, photon emission from TIP modes occurs on clean metals surfaces. Therefore, it is important to further evaluate the role

Fig. 5.26. Photon map of a Au(110) surface covered with a monolayer of C_{60}. The individual molecules appear as bright emission features at a distance of 1 nm. Intensity scale: 800 Hz, temperature of tip and sample: 50 K

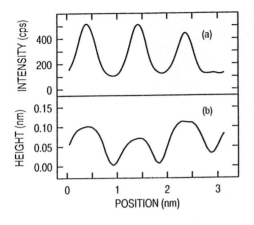

Fig. 5.27. Cross-sectional profiles of photon intensity (a) and topography (b) extracted from Fig. 5.26. When the STM tip is centered above a C_{60} molecule the emission intensity is maximized regardless of bias polarity. Adapted from [5.5]

of the molecule in the photon emission. If the emission from the C_{60}-covered surface were directly due to TIP modes the increase in distance of tip and metal surface caused by the presence of the molecules (typically $\gtrsim 0.4\,\mathrm{nm}$) should weaken the TIP modes. This, in turn, should cause intensity minima above molecules. Such an effect has been discussed above for topographic maxima on Au(110) which were associated with photon intensity minima (Sect. 5.2.4d). A comparison of photon maps and topographs (Fig. 5.27) shows that in contrast to the mechanism reported above for a clean metal surface, *more* intense emission was found *on top* of the molecules. This led to the proposal that the molecules themselves are the primary photon source [5.5]. Coupling between excited molecules or atoms and the plasmons of a metal surface is a well known phenomenon [5.98]. In the unusual geometry of a cavity (tip and sample) the fluorescence may be enhanced in a way that is similar to the mechanisms invoked to explain the surface enhanced Raman (SERS) effect [5.54]. In fact, unusual SERS has recently been reported for C_{60} on Ag lending further support to this idea [5.99]. The dominant role of the molecule itself is also evident from molecularly resolved photon maps of individual anthracene molecules on Ag(110) (Fig. 5.28) [5.100]. The molecular emission is more intense than the emission from the Ag substrate. Moreover, when tunneling to the anthracene, the emission intensity strongly depends on the bias voltage polarity whereas the emission of the surrounding Ag surface is relatively insensitive to it [5.100]. Related measurements of the photon yield from squeezable tunneling junctions with embedded dichloro-anthracene layers recently also have been interpreted using the concept of molecular fluorescence stimulated by inelastic tunneling [5.101].

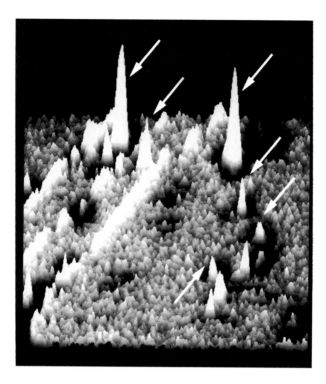

Fig. 5.28. Photon map of isolated anthracene molecules on Ag(110). The molecules appear as bright, "sombrero–shaped" emission features some of which are marked by *arrows*. Temperature of tip and sample: $50\,\mathrm{K}$, $V_t = 2.6\,\mathrm{V}$, $I_t = 30\,\mathrm{nA}$

5.5 Concluding Remarks

To date, STM-induced photon emission is the only scanning probe technique to combine visible light with sub-nanometer lateral resolution. It provides an avenue to investigate inelastic processes that are associated with the tunneling current. The photon intensities encountered from various materials – noble and transition metals, direct-gap semiconductors and molecules on surfaces – permit spatial mapping and photon spectroscopy. The use of electron tunneling as a well-defined and precise distance control as well as the variety of spectroscopic modes available have been instrumental for the experimental characterization and the theoretical interpretation of various factors which determine the photon emission.

From metals, intense emission has been observed and interpreted in terms of tip-induced plasmon modes, which result from the electromagnetic coupling of tip and sample in nanometer proximity. These modes are excited by inelastic tunneling processes. The photon emission provides a direct probe of the intriguing properties of the tip–sample cavity and its coupling to photons. Emission from semiconductors has been analyzed using concepts known from cathodoluminescence. The low electron energies involved permit a lateral resolution in the nanometer range on III–V heterostructures. Local measurements of transport properties and carrier lifetimes as well as band profiling on heterostructures have been demonstrated. More recently exciting observations of new phenomena such as magneto-optical effects and photon emission from individual molecules have been reported. Given the progress that has been made so far for metallic particles, quantum confined semiconductor structures and molecules, STM induced photon emission promises to provide a unique, wide avenue for exploring the diverse optical properties of individual nanoscale objects.

Acknowledgements

I am greatly indebted to J.K. Gimzewski and R.R. Schlittler for sharing with me an exciting research experience during the early studies. It is a pleasure to thank H.J. Güntherodt, H. Rohrer and W.-D. Schneider for their support. Many thanks to A. Baratoff and P. Johansson for discussions on theoretical aspects of photon emission. I am grateful to M. Böhringer, R. Gaisch, B. Reihl, M. Tschudy, and W.-D. Schneider for their contributions to the low-temperature STM studies. Many thanks to J. Beauvillain, S.F. Alvarado and Ch. Bolliger for making available reproductions of figures and K. Ito, S. Ushioda , A. Madrazo, M. Nieto-Vesperinas, N. Garcia and R. Monreal for the permission to use their results prior to publication. I thank E.L. Bullock for helpful discussions and a critical reading of the manuscript.

References

[5.1]   G. Binnig, H. Rohrer: IBM J. Res. Develop. **30**, 355 (1986)
[5.2]   J.K. Gimzewski, B. Reihl, J.H. Coombs, R.R. Schlittler: Z. Phys. B **72**, 497 (1988)
[5.3]   P.K. Hansma: *Tunneling Spectroscopy* (Plenum, New York 1982)
[5.4]   E.L. Wolf: *Principles of Electron Tunneling Spectroscopy* (Oxford Univ. Press, New York 1985)
[5.5]   R. Berndt, R. Gaisch, J.K. Gimzewski, B. Reihl, R.R. Schlittler, W.D. Schneider, M. Tschudy: Science **262**, 1425 (1993)
[5.6]   R. Berndt, R. Gaisch, W.D. Schneider, J.K. Gimzewski, B. Reihl, R.R. Schlittler, M. Tschudy: Phys. Rev. Lett. **74**, 102 (1995)
[5.7]   D.L. Abraham, A. Veider, Ch. Schönenberger, H.P. Meier, D.J. Arent, S.F. Alvarado: Appl. Phys. Lett. **56**, 1564 (1990)
[5.8]   J. Lambe, S.L. McCarthy: Phys. Rev. Lett. **37**, 923 (1976)
[5.9]   D. Hone, B. Mühlschlegel, D.J. Scalapino: Appl. Phys. Lett. **33**, 203 (1978)
[5.10] R.W. Rendell, D.J. Scalapino, B. Mühlschlegel: Phys. Rev. Lett. **41**, 1746 (1978)
[5.11] J.R. Kirtley, T.N. Theis, J.C. Tsang, D.J. DiMaria: Phys. Rev. B **27**, 4601 (1983)
[5.12] R.D. Young, J. Ward, F. Scire: Rev. Sci. Instrum. **43**, 999 (1972)
[5.13] R.D.Young: Phys. Today 42–49 (November 1972)
[5.14] J.K. Gimzewski, J.K. Sass, R.R. Schlittler, J. Schott: Europhys. Lett. **8**, 435 (1989)
[5.15] J.H. Coombs, J.K. Gimzewski, B. Reihl, J.K. Sass, R.R. Schlittler: J. Microsc. **152**, 325 (1988)

[5.16]   N. Venkateswaran, K. Sattler, M. Ge: Surf. Sci. **274**, 199 (1992)

[5.17]   V. Sivel, R. Coratger, F. Ajustron, J. Beauvillain: Phys. Rev. B **45**, 8634 (1992)

[5.18]   M.J. Gallagher, S. Howells, L. Yi, T. Chen, D. Sarid: Surf. Sci. **278**, 270 (1992)

[5.19]   R. Berndt, J.K. Gimzewski, R.R. Schlittler: J. Vac. Sci. Technol. B **9**, 573 (1991)

[5.20]   S. Ushioda, Y. Uehara, M. Kuwahara: Appl. Surf. Sci. **60/61**, 448 (1992)

[5.21]   Recently, conductive transparent tips have been used for low-temperature luminescence studies; T. Murashita: J. Vac. Sci. Technol. B **15**, 32 (1997)

[5.22]   R. Berndt: "Photon Emission from the Scanning Tunneling Microscope"; PhD thesis, University of Basel (1992)

[5.23]   S.F. Alvarado, Ph. Renaud: Phys. Rev. Lett. **68**, 1387 (1992)

[5.24]   K. Takeuchi, Y. Uehara, S. Ushioda, S. Morita: J. Vac. Sci. Technol. B **9**, 557 (1991)

[5.25]   I.I. Smolyaninov, E.V. Moskovets: Phys. Lett. A **165**, 252 (1992)

[5.26]   R. Berndt, A. Baratoff, J.K. Gimzewski: in *Scanning Tunneling Microscopy and Related Methods*, NATO ASI Series E, Vol. 184, ed. by R.J. Behm, N. Garcia, and H. Rohrer (Kluwer, Dordrecht 1990) pp. 269–280

[5.27]   P. Johansson, R. Monreal, P. Apell: Phys. Rev. B **42**, 9210 (1990)

[5.28]   B.N.J. Persson, A. Baratoff: Phys. Rev. Lett. **68**, 3224 (1992)

[5.29]   R. Berndt, J.K. Gimzewski: Phys. Rev. B **45**, 14095 (1992)

[5.30]   R. Berndt, J.K. Gimzewski, P. Johansson: Phys. Rev. Lett. **67**, 3796 (1991)

[5.31]   B.N.J. Persson, A. Baratoff: Bull. Am. Phys. Soc. **35**, 634 (1990)

[5.32]   P. Johansson, R. Monreal: Z. Phys. B **84**, 269 (1991)

[5.33]   P. Johansson: "Theory of Inelastic Tunneling: Applications to Double-Barrier Structures and Scanning Tunneling Microscopes"; PhD thesis, Chalmers University (1991)

[5.34]   P. Johansson, R. Monreal, P. Apell: in *Near Field Optics*, NATO Advanced Studies Institutes Series E, Vol. 242, ed. by D.W. Pohl and D. Courjon (Kluwer, Dordrecht 1993) pp. 341–352

[5.35]   K.R. Welford, J.R. Sambles: J. Mod. Opt. **35**, 1467 (1988)

[5.36]   R. Berndt, J.K. Gimzewski, P. Johansson: Phys. Rev. Lett. **71**, 3493 (1993)

[5.37]   R. Berndt, J.K. Gimzewski: phys. stat. sol. A **131**, 31 (1992)

[5.38]   R. Berndt, J.K. Gimzewski, R.R. Schlittler: Ultramicroscopy **42/44**, 355 (1992)

[5.39]   N. Majlis, A. Levy Yeyati, F. Flores, R. Monreal: Phys. Rev. B **52**, 12505 (1995)

[5.40]   M. Tsukada, T. Shimizu, K. Kobayashi: Ultramicrosc. **42/44**, 360 (1992)

[5.41]   T. Shimizu, K. Kobayashi, M. Tsukada: Appl. Surf. Sci. **60/61**, 454 (1992)

[5.42]   Y. Uehara, Y. Kimura, S. Ushioda, K. Takeuchi: Jpn. J. Appl. Phys. **31**, 2465 (1992)

[5.43]   B. Laks, D.L. Mills: Phys. Rev. B **20**, 4962 (1979)

[5.44]   I.I. Smolyaninov: in *Near Field Optics*, NATO Advanced Studies Institutes Series E, Vol. 242, ed. by D.W. Pohl and D. Courjon (Kluwer, Dordrecht 1993), pp. 353–360

[5.45]   I.I. Smolyaninov, O. Keller: phys. stat. sol. B **185**, 275 (1994)

[5.46]   I.I. Smolyaninov, M.S. Khaikin, V.S. Edelman, Phys. Lett. A **149**, 410 (1990)

[5.47]   I.I. Smolyaninov, V.S. Edelman, V.V. Zavylow: Phys. Lett. A **158**, 337 (1991)

[5.48]   G. Binnig, H. Rohrer: Helv. Phys. Acta **55**, 726 (1982)

[5.49]   K.H. Gundlach: Solid State Electron. **9**, 949 (1966)

[5.50]   A.J. Jason: Phys. Rev. **156**, 266 (1967)

[5.51] G. Binnig, K.H. Frank, H. Fuchs, N. Garcia, B. Reihl, H. Rohrer, F. Salvan, A.R.Williams: Phys. Rev. Lett. **55**, 991 (1985)
[5.52] R.S. Becker, J.A. Golovchenko, B.S. Swartzentruber: Phys. Rev. Lett. **55**, 987 (1985)
[5.53] R. Berndt, J.K. Gimzewski: Ann. Physik **2**, 133 (1993)
[5.54] A. Otto, I. Mrozek, H. Grabhorn, W. Akemann: J. Phys.: Condens. Matter **4**, 1143 (1992)
[5.55] A. Madrazo, M. Nieto-Vesperinas, N. Garcia: Phys. Rev. B **53**, 3654 (1996)
[5.56] A.L. Vasquez de Parga, S.F. Alvarado: Phys. Rev. Lett. **72**, 3726 (1994)
[5.57] A.L. Vasquez de Parga, S.F. Alvarado: Europhys. Lett. **36**, 577 (1996)
[5.58] A.W. McKinnon, M.E. Welland, T.M.H. Wong, J.K. Gimzewski: Phys. Rev. B **48**, 15250 (1993)
[5.59] K. Ito, S. Ohyama, Y. Uehara, S. Ushioda: Surf. Sci. **324**, 282 (1995)
[5.60] T. Umeno, R. Nishitani, A. Kasuya, Y. Nishina: Phys. Rev. B **54**, 13499 (1996)
[5.61] R. Berndt, J.K. Gimzewski: Phys. Rev. B **48**, 4746 (1993)
[5.62] R. Berndt, J.K. Gimzewski: Surf. Sci. **269/270**, 556 (1992)
[5.63] R. Berndt, J.K. Gimzewski: in *Atomic and Nanoscale modification of materials*, NATO Advanced Studies Institutes Series E, Vol. 239, ed. by Ph. Avouris (Kluwer, Dordrecht 1993), pp. 327–335
[5.64] N. Kroo, J.P. Thost, M. Völker, W. Krieger, H. Walther: Europhys. Lett. **15**, 289 (1991)
[5.65] H. Rohrer: in *Scanning Tunneling Microscopy and Related Methods*, NATO ASI Series E, Vol. 184, ed. by R.J. Behm, N. Garcia, and H. Rohrer (Kluwer, Dordrecht 1990), pp. 1–25
[5.66] V. Sivel, R. Coratger, F. Ajustron, J. Beauvillain: Phys. Rev. B **50**, 5628 (1994)
[5.67] V. Sivel, R. Coratger, F. Ajustron, J. Beauvillain: Phys. Rev. B **51**, 14598 (1995)
[5.68] J.K. Gimzewski, R. Berndt, R.R. Schlittler, A.W. McKinnon, M.E. Welland, T. Wong, Ph. Dumas, M. Gu, C. Syrykh, F. Salvan, A. Halimaoui: in *Near Field Optics*, NATO Advanced Studies Institutes Series E, Vol. 242, ed. by D.W. Pohl and D. Courjon (Kluwer, Dordrecht 1993), pp. 333–340
[5.69] S. Ushioda: Sol. Stat. Comm. **84**, 173 (1992)
[5.70] Ph. Dumas, M. Gu, C. Syrykh, J.K. Gimzewski, I. Makarenko, A. Halimaoui, F. Salvan: Europhys. Lett. **23**, 197 (1993)
[5.71] M. Gu, C. Syrykh, Ph. Dumas, F. Salvan: J. Lumin. **57**, 315 (1993)
[5.72] Ph. Dumas, M. Gu, C. Syrykh, A. Halimaoui, F. Salvan, J.K. Gimzewski, R.R. Schlittler: J. Vac. Sci. Technol. B **12**, 2064 (1994)
[5.73] K. Ito, S. Ohyama, Y. Uehara, S. Ushioda: Appl. Phys. Lett. **67**, 2536 (1995)
[5.74] D.L. Abraham, S.F. Alvarado, H.P. Meier, D.J. Arent, Helv. Phys. Acta **63**, 783 (1990)
[5.75] R. Berndt, J.K. Gimzewski, R.R. Schlittler: in *Scanned Probe Microscopies*, Conference Series AIP, Vol. 231, ed. by K. Wickramasinghe (American Institute of Physics, New York 1992), pp. 328–336
[5.76] L. Montelius, F. Owman, M.E. Pistol, L. Samuelson: Inst. Phys. Conf. Ser. Microsc. Semicond. Mater. Conf. **117**, 719 (1991)
[5.77] L. Montelius, M.E. Pistol, L. Samuelson: Ultramicrosc. **42/44**, 210 (1992)
[5.78] L. Samuelson, J. Lindahl, L. Montelius, M.E. Pistol: Phys. Scr. T **42**, 149 (1992)
[5.79] L. Samuelson, A. Gustafsson, J. Lindahl, L. Montelius, M.E. Pistol, J.O. Malm, G. Vermeire, P. Demeester: J. Vac. Sci. Technol. B **12**, 2521 (1994)

[5.80]   S.F. Alvarado: Phys. Rev. Lett. **75**, 513 (1995)
[5.81]   Y. Kuk, P.J. Silverman: J. Vac. Sci. Technol. A **8**, 289 (1990)
[5.82]   R.C. Alig, S. Bloom, C.W. Struck: Phys. Rev. B **22**, 5565 (1980)
[5.83]   F. Steinrisser: Phys. Rev. Lett. **24**, 213 (1970)
[5.84]   M. Wenderoth, M.J. Gregor, R.G. Ulbrich: Sol. Stat. Commun. **83**, 535 (1992)
[5.85]   S.F. Alvarado, Ph. Renaud, H.P. Meier: Journ. Phys. IV **1**, C6 (1991)
[5.86]   O. Albrektsen, D.J. Arent, H.P. Meier, H.W.M. Salemink: Appl. Phys. Lett. **57**, 31 (1990)
[5.87]   Ph. Renaud, S.F. Alvarado: Phys. Rev. B **44**, 6340 (1991)
[5.88]   M. Pfister, M.B. Johnson, S.F. Alvarado, H.W.M. Salemink, U. Marti, D. Martin, F. Morier-Genoud, F.K. Reinhart: Appl. Phys. Lett. **65**, 1168 (1994)
[5.89]   B.G. Yacobi, D.B. Holt: J. Appl. Phys. **59**, R1 (1986)
[5.90]   S.F. Alvarado, Ph. Renaud, D.L. Abraham, Ch. Schönenberger, D.J. Arent, H.P. Meier: J. Vac. Sci. Technol. B **9**, 409 (1991)
[5.91]   M. Bischoff, R. Elm, H. Pagnia, C. Sprössler: Intl. J. Electron. **73**, 1091 (1992)
[5.92]   J. Horn, R. Richter, H.L. Hartnagel, C. Sprössler, M. Bischoff, H. Pagnia: Mater. Sci. Eng. B **20**, 183 (1993)
[5.93]   M. Bischoff, B. Krebs, H. Pagnia, M. Stehle: Intl. J. Electron. **77**, 205 (1994)
[5.94]   M. Stehle, M. Bischoff, H. Pagnia, J. Horn, N. Marx, B.L. Weiss, H.L. Hartnagel: J. Vac. Sci. Technol. B **13**, 305 (1995)
[5.95]   J. Horn, N. Marx, B.L. Weiss, H.L. Hartnagel, M. Stehle, M. Bischoff, H. Pagnia: Materials Science Forum **185/188**, 145 (1995)
[5.96]   R. Berndt, R. Gaisch, J.K. Gimzewski, B. Reihl, R.R. Schlittler, W.D. Schneider, M. Tschudy: Appl. Phys. A **57**, 513 (1993)
[5.97]   R. Berndt, R. Gaisch, W.D. Schneider, J.K. Gimzewski, B. Reihl, R.R. Schlittler, M. Tschudy: Surf. Sci. **307/309**, 1033 (1994)
[5.98]   A. Adams, J. Moreland, P.K. Hansma: Surf. Sci. **111**, 351 (1981)
[5.99]   A. Rosenberg, D.P. DiLella: Chem. Phys. Lett. **233**, 96 (1994)
[5.100] M. Böhringer, R. Berndt: unpublished
[5.101] E. Flaxer, O. Sneh, O. Cheshnovsky: Science **262**, 2012 (1993)

6. Laser Scanning Tunneling Microscope

M. Völcker

With 22 Figures

Why is the coupling of laser light into a scanning tunneling microscope (STM) of interest?

Scanning tunneling microscopy allows the generation of surface images with structures in atomic dimensions. In many cases, the interpretation of these images is problematic. It is often difficult to answer the question: What molecule or atom causes the observed topographic feature? Different particles can lead to identical images. Therefore, additional image information is needed to prove the validity of a microscopic model.

In *light microscopy*, images display not only brightness information, but also color information (and often also polarization). This is more appropiate to the way we see with our eyes. The resolution of optical microscopes is, however, limited to about the wavelength of light. This resolution can be improved by light of a shorter wavelength or by using electrons in electron microscopes. Due to the higher energies involved, these methods can only be used to a limited extent for sensitive samples such as organic material.

Scanning near-field optical microscopy allows the generation of images with brightness, color and polarization information and with spatial resolution well below the wavelength limit mentioned above. Usually, the light has to pass an aperture moved in close proximity to the investigated surface. The spatial resolution achieved with this concept lies in the range of a few tens of nanometers. However, concepts based on other near-field probes may have the potential of attaining atomic resolution in the optical signal [6.1]. One of these is the laser scanning tunneling microscope described in this chapter.

In *optical spectroscopy* many molecular properties are investigated on the basis of frequency analysis alone without spatial resolution. The use of laser light leads to very high spectral resolution, and its high intensities allow the investigation of small effects, for example nonlinear optical processes. Usually, the spectroscopic information results from the interaction of the light with a very large number of atoms. Since different environments of the individual atoms usually shift the individual resonances, there is a high level of interest in obtaining spectroscopic information on single particles.

The *laser scanning tunneling microscope* combines the laser light and its capability to generate images including color information with STM and its capability to generate atomic-resolution images. Here, the light interacts with STM in a localized way. The difficulty of measuring small effects embedded in a large optical background signal can be avoided by detecting, for example, a signal at the difference frequency of two laser frequencies. As discussed later, this difference-frequency signal is generated by local nonlinear effects.

This concept has been followed for several years by a research group at the Max-Planck-Institut für Quantenoptik in Garching close to Munich. Therefore, this chapter is for the most part an overview of the results obtained by this group.

A crucial point in laser scanning tunneling microscopy is to understand the coupling of laser light into the tunneling junction, as well as the interaction of the light with the tunneling junction. For infrared laser light, the coupling is dominated by the antenna properties of the tunneling tip. In the visible spectral range, localized tip-induced plasmons and propagating surface plasmons can be dominant. The laser light coupled into the tunneling junction then induces, on one hand, intensity-dependent effects such as thermal expansion, thermoelectric effects and photoconductivity. On the other hand, field-dependent effects such as the generation of new frequency components are important in addition to the laser frequencies coupled to the STM. The difference-frequency generation is observed to be due to a nonlinear current-voltage characteristic as well as to a nonlinear polarization of the surface. Furthermore, the rectification of the laser field generates a dc currrent, which is observed in addition to the usual tunneling current.

The difference-frequency signal and the rectified current in particular were used in several interesting experiments. It is shown that the images generated by recording these nonlinear signals even display atomic resolution. Moreover, these signals can completely replace the usual tunneling current when they are used for distance control and as an image signal. However, the laser-induced dc current may also cause problems in the usual current-controlled feedback circuit of the tunneling distance. In combination with a scanning force microscope, a laser-induced ac tunneling current is used to study the local conductivity of a sample. In another experiment the laser-induced dc current is used to determine the decay length of propagating surface plasmons. The nonlinear signals also allow measurements with very high time resolution. The most important application of the nonlinear signals is, however, the combination of laser spectroscopy and scanning tunneling microscopy.

6.1 Instrumentation

To avoid the repetition of instrument descriptions, some basic experimental components are sketched at first. Most experiments described in this chapter study the laser-induced dc current or the difference frequency of two laser beams coupled into the junction. Figure 6.1 shows the general setup for these measurements. Laser light of one or two frequencies is focussed in the tunneling junction of an STM. The laser-induced dc current is measured when

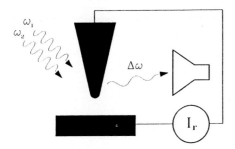

Fig. 6.1. Basic experimental arrangement for laser frequency mixing in a scanning tunneling microscope. Laser light of two frequencies ω_1, ω_2 is coupled into the junction. The rectified current I_r and the radiated signal at the difference frequency $\Delta\omega = \omega_1 - \omega_2$ are measured

no bias voltage is applied between the tip and the sample. The difference-frequency signal is emitted by the tip and collected with a suitable detection system. The following briefly describes the main parts of this setup.

6.1.1 Microscopes

Several microscopes were used for the frequency-mixing experiments. An STM used at ambient air pressure is a small and rigid device. It is easy to handle and allows the imaging of surfaces with atomic resolution. When a cantilever tip is used instead of a tunneling tip, this STM can be combined with a interferometric deflection sensor to build a scanning force microscope. Here a differential configuration similar to the setup of *Schönenberger* and *Alvarado* [6.2] is used. It detects the optical path-length difference of two helium-neon laser beams reflected by the lever. The additional use of a conducting cantilever allows the measurement of the tunneling current, the laser-induced current components, and the force between the tip and the sample at the same time.

For experiments in ultra-high vacuum conditions, an STM was built inside a vacuum chamber. It allows the preparation of wafer-like samples under vacuum conditions and a tip and sample exchange in an easy way. During the measurements the pressure totalled about 2×10^{-10} mbar.

6.1.2 Lasers

For experiments with infrared laser light, two single-mode, line-tunable carbon dioxide lasers are used. Difference frequencies up to 100 MHz are obtained by tuning the lasers across the Doppler-broadened gain profile of the same transition. Using different transitions of different carbon dioxide isotopes, difference frequencies up to 5 THz were measured.

In the near infrared spectral range a tunable color-center laser with a lithium doped potassium chloride crystal provides laser wavelengths of 2.4 to 2.8 μm. Due to spatial hole burning, the laser emits two lines with a difference frequency of 1.35 GHz, which can be used for frequency-mixing experiments. Laser light with a wavelength of 1.064 μm is provided by a Nd:YAG laser.

Experiments in the visible spectral range were performed with a helium-neon laser at 633 nm or with a krypton-ion laser with lines between 413 and 753 nm.

6.1.3 Detection of Nonlinear Signals

The dc tunneling current was measured as usual with a current-voltage converter. Here, however, this current consists of the bias-induced tunneling current and the *laser-induced* dc *current*. This nonlinear signal is measured when no bias voltage is applied between the tip and the sample. For this purpose, the feedback loop is interrupted by means of a sample-and-hold circuit and the bias voltage is switched off for a short period of time. The remaining part of the dc current is only generated due to laser radiation.

The detection system for the laser-induced *difference-frequency signal* depends on the frequency of the signal. In the *megahertz range* the signal is coupled directly out of the tip, amplified, and measured with a spectrum analyzer.

A spectrum analyzer can also be used for difference-frequency signals in the *microwave range*. Also, in order to detect very small nonlinear effects, a very sensitive homodyne-detection system for difference-frequency signals of 9 GHz was built. In this setup, shown schematically in Fig. 6.2, the colinear laser beams of two carbon dioxide lasers are coupled not only into the STM, but also into a Metal–Insulator–Metal (MIM), point-contact diode in order to produce a reference signal at the difference frequency. The difference-frequency radiation generated in the STM and that from the MIM diode are mixed with the output of an oscillator close to 9 GHz at mixer 1 and mixer 2, respectively, in order to convert the laser-induced signals to a lower frequency. The resulting signal of the MIM diode is used to stabilize one laser to a fixed difference frequency with respect to the other free running laser.

For phase-sensitive detection of the difference-frequency signal, the relative phase of the signals of the STM and from the MIM diode can be adjusted

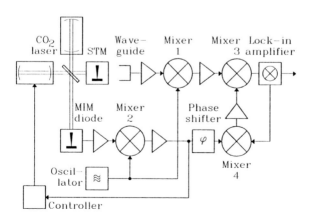

Fig. 6.2. Experimental setup for homodyne detection of a microwave difference-frequency signal [6.3] (slightly modified)

by a phase shifter. Both signals are then bandpass-filtered and superposed at mixer 3. To improve sensitivity, the reference signal is inverted periodically at mixer 4. The resulting modulation of the output is measured with a lock-in amplifier. Depending on its integration time, the homodyne-detection system displays a sensitivity down to about 10^{-21} W.

Difference-frequency signals in the *THz range* were detected by collecting the difference-frequency radiation generated in the tunneling junction with parabolic mirrors and focussing them on a thermal bolometer.

6.2 Coupling of Laser Light into the Tunneling Junction

To study STM in the presence of electromagnetic radiation, laser light has to be coupled into the junction in an effective way. Field enhancement caused by a long-wire antenna geometry, by localized tip-induced plasmons and by propagating surface plasmons are discussed in the following.

For infrared light the very end of the tip is thinner than the wavelength so that it serves as a *long-wire antenna*. A conducting sample serves here as a reflector, doubling the effective antenna length. These antenna properties have also been observed for MIM diodes [6.4]. Figure 6.3 shows the basic arrangement for this way of coupling laser light into the tunneling junction. Here, the angle of incidence between the tip axis and the laser beam is typically choosen to be about 30°, which is approximately the angle of the main lobe of the antenna pattern (Fig. 6.4). For effective induction of antenna currents, the polarization direction has to be parallel to the plane of incidence.

Due to these antenna properties antenna currents flow at the tip, and oscillate at the laser frequency. They induce charges at the very end of the tip and mirror charges in the conducting sample. These charges lead to an oscillating voltage, in addition to the bias voltage applied, and to an oscillating electric field between the tip and the sample. The laser-induced electric and magnetic fields of a conical antenna are depicted in Fig. 6.5, based on

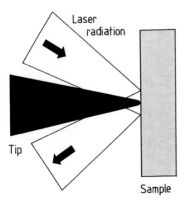

Fig. **6.3.** Basic arrangement for coupling of laser light via antenna or via tip-induced plasmons [6.5]

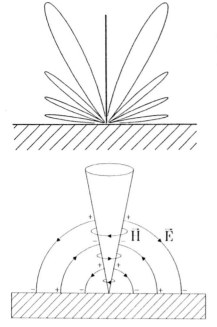

Fig. 6.4. Antenna pattern of a long-wire antenna. The example shows the polar plot of the radiated or received field of a wire, four wavelengths long, perpendicular on an ideal reflector [6.5]

Fig. 6.5. Schematic drawing of the electromagnetic fields E and H of a conical antenna [6.5]

the antenna theory of *Sullivan* et al. [6.6] and *Schelkunoff* [6.7]. A detailed calculation of the laser-induced antenna currents and the resulting oscillating voltages are given in [6.5, 3]. Due to the geometry present, the electric field between the tip and the sample is much larger than the incident laser field. This tip-induced field enhancement is calculated to be in the order of 1000, depending on the radius of curvature of the tip and the tip-sample distance [6.6].

For visible laser light the tip is not sharp enough to serve as an antenna. However, in this spectral range laser radiation can be coupled into the tunneling junction by means of tip-induced plasmon oscillations or by means of propagating surface plasmons. To excite *tip-induced plasmons*, the laser light is focussed in the tunneling junction, as displayed in Fig. 6.3. A typical mode of this excitation is indicated in Fig. 6.6, and calculations for this geometry were performed by *Johansson* et al. [6.9].

The Kretschman geometry is used to excite *propagating surface plasmons* (Fig. 6.7). In this setup the laser light is totally reflected from the metal-coated side of a dielectric prism. For a certain angle of incidence, the surface plasmons are excited via attenuated total reflection (for details see [6.10, 11]). Figure 6.8 shows then the electric field of surface plasmons propagating parallel to the surface as indicated by the momentum vector k_0.

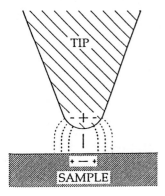

Fig. 6.6. Schematic drawing of a tip-induced plasmon mode. The energetically lowest mode of the electromagnetic resonance, which is approximately dipolar, is shown [6.8]

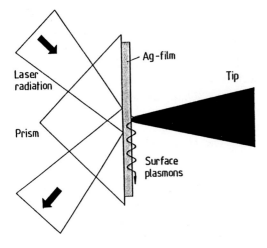

Fig. 6.7. Basic arrangement for coupling light into an STM via excitation of propagating surface plasmons [6.5]

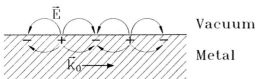

Fig. 6.8. Schematic drawing of the charge distribution and the electrical field E of propagating surface plasmons with k_0 as plasmon momentum at the vacuum–metal interface [6.5]

6.3 Interaction of the Tunneling Junction with Laser Light

6.3.1 Thermal Effects

The interaction of the tunneling junction with laser light is always accompanied by *heating*. The heat transfer from a heated tip to the sample was investigated by *Williams* and *Wickramasinghe* [6.12] using a miniaturized thermocouple instead of the usual tunneling tip. *Wickramasinghe* [6.13] wrote a review on this field.

The focussing of laser light into the tunneling junction results in heating of both the tip and the sample. The temperature rise of the tip and the sample results in thermal expansion of the irradiated parts [6.14]. This has to be compensated by a contraction of the piezo translator controlling the tip-sample distance. Therefore, the laser power has to be kept constant when the tip is close to the surface. The time response of the tip and the sample to modulated laser light was investigated by *Grafström* et al. [6.15]. The modulation of the tunneling current caused by a thermally modulated tip-sample distance was used by *Probst* et al. [6.16] to measure the apparent barrier height of an octylcyanobiphenyl film adsorbed on graphite.

An additional consequence of the temperature rise in the tunneling junction is the generation of *thermoelectric currents*. These can be produced, on the one hand, by an elevated temperature of the tip and the sample representing a thermocouple consisting of different materials [6.17]. *Williams* and *Wickramasinghe* [6.18] obtained surface images with atomic resolution by periodically measuring the thermoelectric voltage while the feedback circuit is interrupted and the tunneling current is switched off. On the other hand, thermoelectric currents can be produced by a higher tip temperature, even when the two materials of the tip and the sample are identical [6.19].

6.3.2 Photovoltaic Effects

Using semiconducting samples, the interaction of the tunneling junction with laser light results in photoexcitation of electrons across the band gap. This leads to an increase in surface conductivity, which was used by *van de Walle* et al. [6.20, 21] to permit scanning tunneling microscopy on semi-insulating gallium-arsenide at room temperature.

In addition, the photoexcitation of electron-hole pairs leads to a surface photovoltage, as the electron-hole pairs are separated due to the electric field of the semiconductor space-charge layer [6.22]. By recording this laser-induced voltage as a function of tip location, it is possible to obtain a spatially resolved map of the surface photovoltage. The contrast mechanism in these measurements is controversial. As summarized by *McEllistrem* et al. [6.23] the proposed contrast mechanisms include spatial variations in the local surface recombination rate [6.22], optical rectification [6.24], experimental artifacts [6.25], charging effects [6.26], and tip-induced band bending [6.23].

6.3.3 Generation of Nonlinear Signals

The generation of new frequency components (e.g., difference-frequency signals, harmonics or rectified currents) in addition to the laser frequencies coupled into the STM requires intrinsic nonlinearities of the microscope. One of these is the quadratic dependence of the thermal expansion of the tip and the sample on the laser field, resulting in modulation of the gap width at the difference frequency. Furthermore, the nonlinear dependence of the tunneling current on the gap width leads to the generation of harmonics of the difference-frequency signal. Also, the nonlinearity due to the current-voltage characteristic of the difference junction generates the difference-frequency signal and rectified current because laser radiation induces an oscillating voltage between the tip and the sample. Mixing can also be caused by a nonlinear polarization of the sample surface.

The *nonlinearity due to thermal modulation* of the tunneling gap is the dominant mixing mechanism in the kilohertz and megahertz range. As an example, the frequency dependence of the difference-frequency signal is plotted in Fig. 6.9. The measured signals were corrected for a reduction of the laser intensity at higher difference frequencies. As seen in the figure, the signal amplitude decreases with increasing difference frequency. The measured points follow a straight line in a double logarithmic plot corresponding to an inverse difference-frequency dependence.

The difference-frequency signals are produced by temperature modulation of the tip at the difference frequency. This temperature modulation modifies the gap width due to thermal expansion. The observed inverse difference-frequency dependence of the mixing signals is then explained by assuming a thermal time constant for the front end of the tip given by the heat capacity of its irradiated part and the heat conductance through its shank [6.28]. This time constant is estimated to be $100\,\mu s$ which is consistent with the results shown in Fig. 6.9.

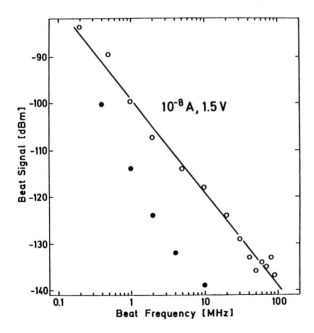

Fig. 6.9. Plot of the mixing (beat) signal vs mixing (beat) frequency in the megahertz range: o difference-frequency signal. The *straight line* indicates an inverse frequency dependence. • second-harmonic signal of the difference frequency [6.27]

Second- and third-harmonic signals of the difference frequency were also recorded. The second-harmonic signals, which are plotted as full circles in Fig. 6.9, were observed up to a frequency of 10 MHz, where the noise limit was reached. Third harmonics were obtained at difference frequencies of 1 MHz and below. Higher harmonics of the difference frequency are generated by the nonlinear dependence of the tunneling current on the gap width.

For difference-frequency signals in the gigahertz range, the main contribution to mixing is the *nonlinearity caused by the current-voltage characteristic* of the tunneling junction. Provided the response of the junction is sufficiently fast, the static current-voltage characteristic can be used to calculate the new current components. The total tunneling current I can be expressed as a power series in terms of the laser-induced voltage V_i:

$$I = I(V_b) + \left.\frac{\partial I}{\partial V}\right|_{V_b} V_i + \frac{1}{2}\left.\frac{\partial^2 I}{\partial V^2}\right|_{V_b} V_i^2 + \dots , \tag{6.1}$$

where V_b is the external bias voltage, and the derivatives are evaluated at V_b. The rectified current results in

$$I_R(V_b) \sim \left.\frac{\partial^2 I}{\partial V^2}\right|_{V_b} \sqrt{P_{\omega 1}P_{\omega 2}} , \tag{6.2}$$

where $P_{\omega 1}$ and $P_{\omega 2}$ are the laser powers at ω_1 and ω_2. As the radiated power $P_{\Delta\omega}$ at the difference frequency $\Delta\omega = \omega_1 - \omega_2$ grows as $I_{\Delta\omega}^2$, one obtains

$$P_{\Delta\omega}(V_b) \sim \left(\left.\frac{\partial^2 I}{\partial V^2}\right|_{V_b}\right)^2 P_{\omega 1}P_{\omega 2} . \tag{6.3}$$

A characterization of the STM junction as a frequency mixer is obtained by a simultaneous measurement of the current-voltage characteristic and the

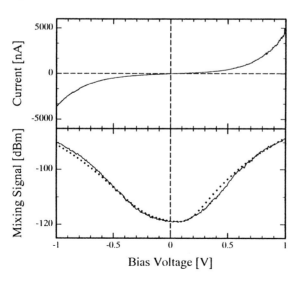

Fig. 6.10. Plot of the tunneling current (*upper part*) and the mixing signal at 9 GHz (*lower part*) vs tip bias voltage. The *dotted curve* indicates the calculated mixing signal [6.29]

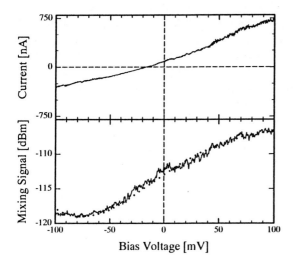

Fig. 6.11. Plot of the tunneling current (*upper part*) and the mixing signal at 9 GHz (*lower part*) vs tip bias voltage. The *dotted curve* indicates the calculated mixing signal. At zero bias voltage the rectified current can be seen [6.29]

bias dependence of the mixing signal. An example is presented in Fig. 6.10. Similar measurements are performed for different values of the preset feedback loop current. Decreasing the tip-sample distance leads to an increase of the mixing signal. A detailed discussion of these results is presented in [6.30]. The rectified current prevents the current-voltage characteristic from passing through the origin of the diagram in Fig. 6.11.

Signals were observed at difference frequencies up to about 5 THz, which is the maximum available difference frequency with the carbon dioxide lasers. As an example, the dependence of the mixing signal on the applied laser power at a difference frequency of 1 THz is shown in Fig. 6.12.

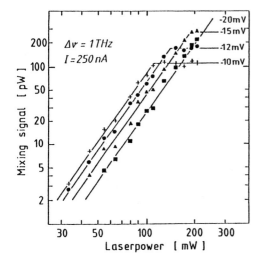

Fig. 6.12. Difference-frequency signal at 1.07 THz vs laser power at four different bias voltages. The *solid line* serves as a guide for the eye [6.31] (slightly modified)

A further mixing mechanism is the generation of a nonlinear displacement current due to a *nonlinear polarization* at the surface. This mixing mechanism therefore corresponds to a nonlinear optical process. Many mechanisms have been proposed to account for the nonlinear polarization in nonlinear optics. Here, the nonlinear surface polarization is explained by surface electrons oscillating in the anharmonic surface potential, driven by the tip-enhanced electrical field of the lasers. In addition, the rapid spatial variation of the electrical field in the skin depth creates nonlinear polarization. These two contributions to the surface nonlinearity can be described by an effective nonlinear susceptibility $\chi_S^{(2)}$ of the surface [6.32].

The nonlinear susceptibility gives rise to a nonlinear displacement current. For the case of two laser frequencies ω_1 and ω_2 coupled into the difference junction, the nonlinear displacement current includes a term oscillating at the difference frequency [6.33]

$$I_{\Delta\omega} \sim -\frac{\chi_S^{(2)}\sqrt{P_{\omega 1}P_{\omega 2}}}{a^2} \, , \tag{6.4}$$

where $P_{\omega 1}$ and $P_{\omega 2}$ are the laser powers at ω_1 and ω_2, and $a = R + d$ consists of the radius of curvature R of the STM tip and the tip-sample distance d.

As can be seen in (6.4), the nonlinear displacement current has a distance dependence proportional to $1/a^2$, which is much weaker than the exponential distance dependence of the usual tunneling current and the laser-induced ac or dc tunneling current components. The difference-frequency signals caused by a nonlinear displacement current and by a laser-induced ac tunneling current can therefore be distinguished by their different dependence on the tip-sample distance.

Figure 6.13 shows a plot of the dependence of the tunneling current, the laser-induced dc current and the difference-frequency voltage at 9 GHz on the tip-sample distance. Both the tunneling current and the laser-induced dc current decrease exponentially within a few nanometers. This corresponds to

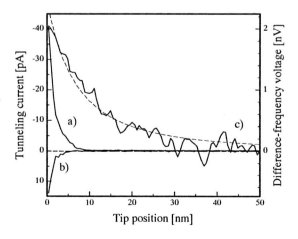

Fig. 6.13. Distance dependence of (*a*) the tunneling current, (*b*) the laser-induced dc current, and (*c*) the difference-frequency signal at 9 GHz. The *dashed line* is a theoretical curve as described in the text [6.33]

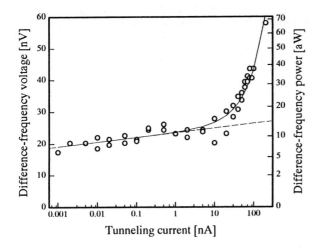

Fig. 6.14. Difference-frequency signal vs tunneling current. The *lines* are theoretical curves representing the total difference-frequency signal (*solid line*) and the part of the signal which is generated due to a nonlinear displacement current (*dashed line*) [6.33]

the model described above because the nonlinear displacement current has no dc component. The difference-frequency voltage decreases less steeply and is still detectable at larger distances. It obeys the distance dependence of (6.4). A curve proportional to $1/a^2$ fitted to the measured data is shown as a dashed line in the figure.

A plot of the difference-frequency signal vs the tunneling current is presented in Fig. 6.14. The tip–sample distance was kept constant by the feedback control during the measurement of one single point. The different distance dependences of the two mechanisms for difference-frequency generation cause the twofold behavior of the plot. For tunneling currents larger than about 1 nA, the main part of the difference-frequency signal is due to mixing at the nonlinearity of the current-voltage characteristic, while for tunneling currents less than about 1 nA the dominant part is generated by a nonlinear displacement current.

6.3.4 Microscopic Model

A model describing the laser-induced currents in the STM was developed by *Yeyati* and *Flores* [6.34–36]. A tight-binding description of the electronic structure of the microscope tunneling junction is used. The interaction with the laser field is taken into account by a time-dependent coupling between the tip and the sample. The difference-frequency signal and the rectified current are then calculated in terms of nonequilibrium Green's functions. The results are applied to graphite samples.

The model predicts a maximum in the rectified current for a tip-sample distance between 0.1–0.2 nm away from close contact [6.34]. A similar behavior with a maximum and a rapid falloff with increasing tip-sample distance was also found in calculations of *Miskovsky* et al. [6.37]. In the limit of low frequencies the dynamic response of the STM may be deduced from the static current-voltage characteristic [6.36]. This limit corresponds to the model described in Sect. 6.3.3. The rectified current is also calculated as a function of lateral displacement. The maxima correspond to the center of the hexagons in the graphite lattice rather than to the atomic positions [6.35]. This result is compared with experimental results in Sect. 6.5.1.

6.3.5 Distance Control

If thermoelectric effects are neglected, the dc tunneling current in an STM with laser irradiation still consists of two components:

$$I_b + I_r(P_1) \ . \tag{6.5}$$

Here, I_b is the tunneling current caused by the externally applied bias voltage, and I_r is the laser-induced dc current which is proportional to the laser power P_1. At small bias voltages and large laser powers, I_r is observed to be of the same order of magnitude as I_b. As an important consequence of this raise, the tunneling distance in the usual current-controlled STM is crucially changed by additionally generated current components. If I_r and I_b flow in the same direction, the tunneling gap will be increased with growing laser power. As the nonlinear current components become smaller at increasing tunneling distance, this leads to a limitation of the mixing signal with increasing laser power. If I_b and I_r are of opposite sign, the tunneling gap is decreased with growing laser power, thereby increasing the nonlinear current components to a greater extent than the expected laser power dependences according to (6.2 and 3). In this case the tunneling distance becomes much smaller than given by the preset tunneling parameters without laser irradiation.

6.4 Applications of Photovoltaic Effects

6.4.1 Time Resolution

The speed limitations conventionally encountered in scanning probe microscopy result from the external electronics and are not inherent to the techniques themselves. A time resolution faster than the bandwidth of the measuring electronics can be achieved by combining these techniques with picosecond optical excitation and utilizing nonlinearities in the physical system. *Hamers* and *Cahill* [6.38] applied pulsed laser radiation to silicon surfaces, determining the relaxation time of carriers on the nanosecond time scale via measurements of the surface photovoltage (Sect. 6.3.2). Therefore, the nonlinear dependence of the surface photovoltage on the illumination intensity makes the time-averaged surface photovoltage sensitive to the temporal distribution of optical pulses.

6.4.2 Photo-Induced Tunneling Current

Akari et al. [6.39] measured the photo-induced tunneling current as a function of wavelength. This provides local spectroscopic information on the photon-electron interaction of semiconducting materials. On a p-type semiconducting tungsten-diselenide a large photo-induced tunneling current with a positive voltage at the tip was observed [6.40]. This current starts increasing below a wavelength corresponding to the bandgap. It shows, as a function of wavelength, a number of small dips propably caused by excitons [6.39].

6.5 Applications of Nonlinear Signals

6.5.1 Laser-Driven Scanning Tunneling Microscope

The following experiments use the spatial resolution of the STM and demonstrate that surface images can be obtained not only by recording the tunneling current, but also by recording the difference-frequency signal or the rectified current of the laser radiation.

Figure 6.15a exhibits a schematic drawing of the experimental setup for the simultaneous recording of graphite surface images using the tunneling current and the difference-frequency signal at 9 GHz. The corresponding images are displayed in Fig. 6.15b and c, respectively. The maxima (bright spots) of the tunneling current and of the difference-frequency signal occur in this example at different locations on the graphite surface, as can be recognized by means of the two intersecting lines. Similar images are obtained by recording the tunneling current and the rectified current [6.42].

Since both the difference-frequency signal and the rectified current increase with decreasing tunneling distance and display atomic resolution, both signals can completely replace the usual tunneling current in the feedback circuit. This means, they can be used to control the tip-sample distance and to generate surface images.

As shown in Fig. 6.16a, the rectified current is used to control the tip-sample distance in the STM. Since in this setup the laser generates a current in the STM, no externally applied bias voltage is necessary. An image of the graphite surface obtained in this mode of operation by monitoring the rectified signal is displayed in Fig. 6.16b. As in Figs. 6.15b and c, high values of the current correspond to the bright spots in the figure.

In a further experiment the intensity of the 9 GHz difference-frequency signal controls the tip-sample distance and generates surface images. As the emitted radiation is used for distance control as well as for imaging, no external bias voltage and no dc current measurement are necessary. All electrical connections to the sample can therefore be removed. A schematic drawing of the experimental setup is presented in Fig. 6.17a. In this circuit the output of the spectrum analyzer is connected to the feedback control. The distance

(a)

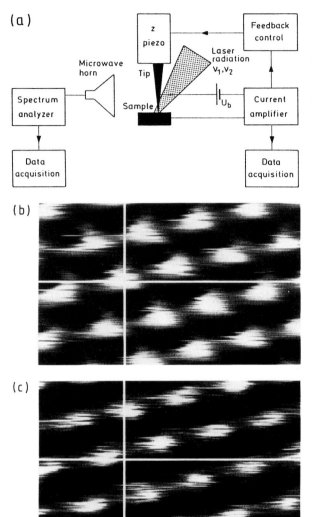

Fig. 6.15a–c. Atomic resolution with the difference-frequency signal. (a) Experimental setup for the generation and detection of laser difference frequencies (U_b: bias voltage). (b, c) Images of a graphite surface obtained by simultaneously recording the tunneling current (b) and the difference-frequency signal (c) [6.41]

(b)

(c)

is adjusted so that the average signal is kept at a preset level. With this configuration, images of the graphite surface are obtained by recording the intensity of the mixing signal. An example is displayed in Fig. 6.17b.

The mode of operation with laser-induced signals used for the generation of surface images may also be interpreted as scanning near-field optical microscopy, as the geometry of the tunneling junction enhances the laser field in an area much smaller than the wavelength. In comparison with the usual near-field probes based on aperture concepts, however, the intensity at the tip end is much higher. Here, the possibility of obtaining atomically resolved images, which has not yet been achieved with other near-field optical

(a)

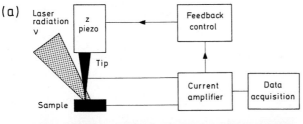

(b)

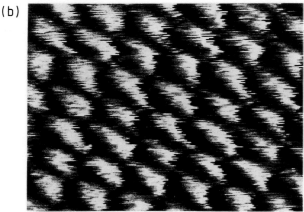

Fig. 6.16. (a) Setup of a laser-driven STM. The tip-sample distance is controlled by the rectified current. Note the absence of a bias voltage source in the tunneling circuit. (b) Image of a graphite surface obtained by recording the rectified current [6.41]

(a)

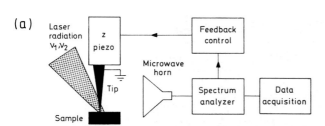

(b)

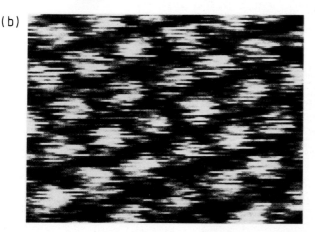

Fig. 6.17. (a) Setup of a laser-driven STM. The tip-sample distance is controlled by the difference-frequency signal. Note the absence of any electrical connection to the sample. (b) Image of a graphite surface obtained with this setup by recording the difference-frequency signal [6.41]

methods [6.1], is very attractive. However, the close connection to the tunneling current limits applicability to conducting samples.

Since no resonant processes are involved here the laser frequencies can be tuned. Therefore, this method also has the capability of optical spectroscopy, which is the advantage of scanning near-field optical microscopy compared to standard scanning tunneling microscopy.

6.5.2 Local Conductivity

Recording the difference-frequency signal allows the generation of surface images of graphite without electrical connection to the sample, as described in Sect. 6.5.1. Can this method be applied to samples with only locally conducting parts, such as wires on a chip or clusters on nonconducting surfaces?

A pattern of gold islands of various sizes on a nonconducting barium-difluoride substrate serves as a test structure for answering this question. A photograph taken with an optical microscope is depicted in Fig. 6.18b, and the experimental arrangement in Fig. 6.18a. Images generated by means of the difference-frequency signal are recorded simultaneously with the topography. This signal is detected with the homodyne-detection system, as described in Sect. 6.1.3. The tip-sample distance is controlled by the force. A large field enhancement in the tunneling gap is achieved by a tip acting as a long-wire antenna for the laser field (Sect. 6.2). This tip was made by modifying

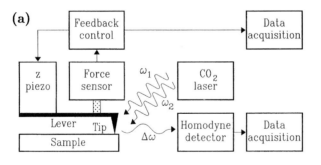

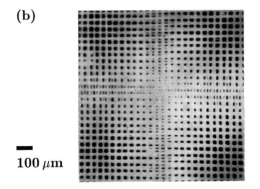

100 μm

Fig. 6.18. (a) Experimental setup for the investigation of samples with only local conducting parts. (b) Photograph of a test pattern consisting of gold islands (*dark*) on nonconducting barium-difluoride [6.43]

(a)

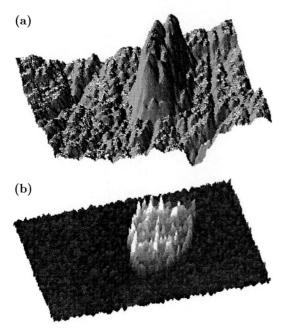

(b)

Fig. 6.19. (a) Topography and (b) image obtained by recording the difference-frequency signal at 9 GHz. The large hill in the center of (a) corresponds to a gold island. The area displayed is $10 \times 6.5\,\mu m^2$ [6.3]

commercial platinum-iridium STM tips so that it also acts as a scanning force microscope lever [6.43].

Figure 6.19a displays a topographic image obtained with the force microscope. The area displayed is $10 \times 6.5\,\mu m^2$, with a maximum height difference of 140 nm. In the center a large hill corresponds to an evaporated gold island. The image shown in Fig. 6.19b is generated from the simultaneously recorded difference-frequency signal. A difference-frequency signal is only observed on the gold island.

When different islands on the test pattern are examined, a strong dependence of the difference-frequency signal on the island size was found, as shown in Fig. 6.20. For an explanation of the sharp drop of the signal with decreasing island size, the charges on the tip-sample geometry are considered (Fig. 6.21). When the island size becomes comparable to the wavelength, a reduction of island size will strongly decrease the charges close to the junction and hence the difference-frequency signal as well.

In reply to the question posed at the beginning, the experiment shows that conducting and nonconducting parts of a surface can be distinguished by observing the difference-frequency signal. The signal drops with decreasing island size is inherent in this method, since the coupling efficiency of laser radiation into the difference junction is reduced for islands smaller than the laser wavelength. Using laser radiation with shorter wavelengths may allow the investigation of smaller islands.

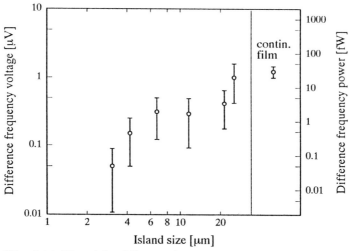

Fig. 6.20. Plot of the difference-frequency signal vs the island size. For comparison, the difference-frequency signal on a continuous gold film is plotted on the right [6.43]

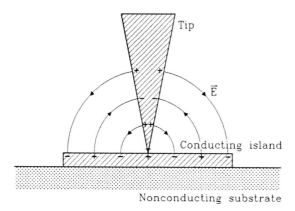

Fig. 6.21. Schematic drawing of an instantaneous distribution of the oscillating charges on the tip, of the corresponding mirror charges on the conducting island, and of the electric-field lines connecting them [6.43]

6.5.3 Propagating Surface Plasmons

The laser-induced dc signals are used to investigate propagating surface plasmons. These plasmons are excited on thin metal films by using the method of attenuated total reflection as sketched in Fig. 6.7. Two mechanisms were identified to contribute to the plasmon-induced dc signal in the tunneling current: the thermal expansion of the tip and the sample, and the rectification of the plasmon field due to the nonlinearity of the current-voltage characteristic of the junction [6.44, 45]. The comparison of the current induced by the plasmon field ($\approx 10^{14}$ Hz) with low frequency fields (10^3 Hz) measured in UHV conditions supports the model of rectification [6.46]. The amplitude of the plasmon-field depends crucially on the tip material and is enhanced by a factor of several hundreds for a silver tip [6.46]. This field enhancement is due to localized plasmons, as described in Sect. 6.2.

Möller et al. [6.45] mapped the local varitions of the plasmon-induced signal while scanning the tip. These experiments have been performed in air, hence the current-voltage characteristic can be completely changed by impurities.

Kroo et al. [6.44] determined the decay length of propagating surface plasmons by recording the plasmon-induced signal during scanning of the laser focus, which is kept on the gold film. In this arrangement local variations in the plasmon-induced signal can be avoided by keeping the tip position fixed. Here the thermal-expansion signal is used to determine the plasmon decay from a line scan [6.44] or from a two dimensional map [6.47].

In the latter experiment the beam of a cw Nd:YAG laser at 1064 nm is focussed by a lens through the prism to excite surface plasmons on a gold film. The lateral distance between the tunneling junction and the laser focal spot is varied by moving the lens. An image is generated by recording the plasmon-induced signal during scanning of laser focus. The tunneling current is kept constant by the feedback loop in the usual way. The plasmon-induced change of the tunneling current is determined by using an intensity-modulated laser beam and by measuring the current with a boxcar averager. As an example, an image of the plasmon-induced signal excited on a 40 nm gold film is presented in Fig. 6.22. From the signal decrease, the plasmon decay length can be determined to be 84 μm.

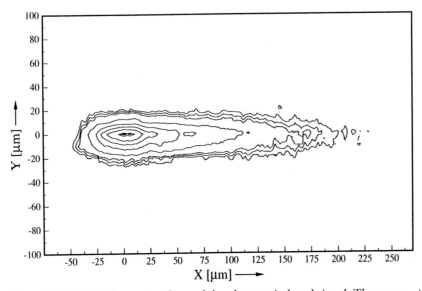

Fig. 6.22. Plot of the contour lines of the plasmon-induced signal. The propagating surface plasmons are excited by a laser spot located at the origin of the plot. The plasmons are propagating to the right [6.47]

6.5.4 AC Scanning Tunneling Microscope

Kochanski [6.48] was the first to apply a microwave voltage to the tunneling gap and study the generation of higher harmonics of the tunneling current. He used the generated harmonics to control the feedback loop of an STM. Thus, he was able to scan insulating films and semiconductors without any dc tunneling current. The physical mechanism for these interesting observations remains unclear.

Michel et al. [6.49] measured higher-harmonic signals when radio frequency fields were applied between the tip and the sample. The close association of the third-harmonic signal and the rectified current indicates that the nonlinear current-voltage characteristic of graphite is the source of the nonlinear effects. Similar results were observed by *Seifert* et al. [6.50]. On silicon the nonlinear current-voltage characteristic is not significant for the third-harmonic signal [6.49]. Instead, it is believed that the signal is caused by a nonlinear polarization of the surface. For both samples images are recorded where the third-harmonic signal is used to control the tip-sample distance. Images of graphite show local changes of the nonlinearity, for example atomic steps or domain boundaries. Images of silicon with an oxide layer show a change of the nonlinear polarization.

In the case of semiconducting tungsten-diselenide, *Seifert* et al. [6.50] attributed the origin of the observed generation of higher harmonics to the field-induced change of the space-charge capacitance near the semiconducting surface.

Stranick et al. [6.51] applied tunable microwave frequencies to an STM which can be operated at low temperatures and in ultra-high vacuum. In contrast to the microwave experiments mentioned above no resonant microwave cavity was used. By controling the tip-sample distance with the microwave signal, atomic resolution on graphite and tungsten-diselenide was achieved [6.52]. On insulating surfaces such as lead silicate glass the resolution was approximately 1 nm.

6.5.5 Tunneling Time

Laser illumination of an STM tip modulates the tunneling barrier by superimposing the electric field of light. If the junction is irradiated by a laser of fixed frequency where the polarization is parallel to the axis of the tip, the electron transfer process is enhanced during half of the period when the the electric-field vector of the laser beam is accelerating. During the second half of the period when the laser field is reversed, the barrier will increase and transport will be diminished [6.53]. If the gap is larger than a certain critical value s_c, the electron will not have "enough time" to transverse the barrier and to be detected before the field is reversed. The tunneling time can be deduced from $\tau = s_c/v$, where it is found that $v \approx v_F$, the Fermi velocity of the electrons in the metallic electrodes [6.37].

The dependence of tunneling current on the laser frequency can serve as a basis for determining the duration of barrier traversal. The barrier modulation by light induces a rectified current, but this effect disappears at frequencies much greater than the reciprocal of the tranversal time [6.54].

An operational tunneling time was determined by *Nguyen* et al. [6.55] who measured the rectified current in a laser-illuminated STM. In this experiment, a linearly polarized Nd:YAG laser of wavelength 1.06 μm was focused on an STM junction consisting of a tungsten tip and a silicon base in ultra-high vacuum. They deduced a tunneling time of about 1.8 fs. However, the tunneling time is still open to interpretation [6.37]. The observed decrease in rectified current with increasing barrier length is similar to the exponential dependence of transmission on the barrier length so that this data may be explained without referring to tunneling times [6.54].

In a similar experiment *Möller* et al. [6.56] measured the distance dependence of the plasmon-induced signal and of the tunneling current. Thermal heating was excluded from the measurement by using repeated short optical pulses, thereby utilizing the fact that thermal relaxation times are longer than nonthermal effects of illumination.

Regarding the frequency limits for the dynamical response in STM, the model of *Yeyati* and *Flores* [6.36] suggests that there is no clear frequency cut-off except for photon energies comparable to the total electronic band widths. Obviously, the assumption of perfect screening for the coupling of the electromagnetic field no longer applies at lower frequencies in the region of surface plasmons. Thus, it is not the electron tunneling time but rather the electromagnetic coupling which limits the dynamic response [6.36].

6.6 Spectroscopy

The most important application of the laser STM lies in the combination of laser spectroscopy and scanning tunneling microscopy in order to obtain information on the substrate under study with spectroscopic and high spatial resolution. Many different ways are proposed to achieve this goal. In the following, some of them are summarized.

Akari et al. [6.39] used light of a tungsten-halogen lamp for illumination of an STM and recorded the photo-induced tunneling current as a function of wavelength (Sect. 6.4.2).

The generation of laser-induced nonlinear signals opens up the possibility for obtaining spectroscopic information. On a surface with adsorbed molecules difference-frequency generation caused by a nonlinear polarization of the surface is expected to be strongly enhanced if a molecular resonance, e.g., vibration excitation in the infrared spectral range, is involved in the frequency mixing process [6.33]. The mixing mechanism corresponds to frequency mixing in nonlinear optics such as second-harmonic or sum-frequency generation, with the advantage that the obtainable lateral resolution is considerably increased.

In the microwave range *Michel* at al. [6.49] proposed the use of higher-harmonic generation. Preliminary results indicate that it is possible to identify molecular adsorbates on gold by their larger third-harmonic signal.

Another experiment in the microwave range was proposed by *Stranick* et al. [6.51]: By monitoring the ac components of the tunneling current, the authors expect to be able to measure rotational frequencies of single adsorbed molecules. It should also be possible to identify and characterize these molecules by investigating the amplitude of the harmonics of a microwave modulation frequency.

The use of frequency mixing due to a nonlinear current-voltage characteristic was considered theoretically by *Yeyati* and *Flores* [6.36]. For this purpose, the model described in Sect. 6.3.4 was generalized so that the effect of an adsorbed molecule on the laser-induced currents is included [6.36]. In this theory the rectified current is calculated in the presence of an adsorbed molecule whose vibrational mode is excited by laser light. The contribution from inelastic tunneling processes is obtained to the lowest order in the electron-phonon coupling. It is shown that the onset of inelastic tunneling should be reflected as a singularity in the rectified current as a function of both bias voltage and phonon energy. Information on molecule resonant level, its width and position relative to the Fermi energy, and the coupling between electrons and localized vibrations can be obtained from this signal.

Nevertheless, a convincing experimental demonstration of an STM combined with laser spectroscopy is still the outstanding challenge of the future.

Acknowledgements

The author would like to thank C. Sammet for her very helpful assistance and W. Krieger and C. Poole for careful reading of the manuscript.

References

[6.1] O. Marti, G. Krausch: Phys. Bl. **51**, 493 (1995)
[6.2] C. Schönenberger, S.F. Alvarado: Rev. Sci. Instrum. **60**, 3131 (1989)
[6.3] M. Völcker, W. Krieger, H. Walther: J. Vac. Sci. Technol. B **12**, 2129 (1994)
[6.4] Bor-Iong Twu, S.E. Schwarz: Appl. Phys. Lett. **26**, 672 (1975)
[6.5] M. Völcker: Rastertunnelmikroskopie mit Einstrahlung von Laserlicht. Ph.D. Thesis, Ludwig-Maximilians-Universität München (1992)
[6.6] T.E. Sullivan, P.H. Cutler, A.A. Lucas: Surf. Sci. **62**, 455 (1977)
[6.7] S.A. Schelkunoff: *Electromagnetic Waves* (van Nostrand, New York 1943)
[6.8] R. Berndt, R. Gaisch, W.D. Schneider, J.K. Gimzewski, B. Reihl, R.R. Schlittler, M. Tschudy: Appl. Phys. A **57**, 513 (1993)
[6.9] P. Johansson, R. Monreal, P. Apell: Phys. Rev. B **42**, 9210 (1990)

[6.10] E. Kretschmann, H. Raether: Z. Naturforsch. **23a**, 2135 (1968)
[6.11] H. Raether: *Surface Plasmons* (Springer, Berlin, Heidelberg 1988)
[6.12] C.C. Williams, H.K. Wickramasinghe: Appl. Phys. Lett. **49**, 1587 (1986)
[6.13] H.K. Wickramasinghe: In *Scanning Tunneling Microscopy II*, 2nd edn., ed. by R. Wiesendanger, H.-J. Güntherodt, Springer Ser. Surf. Sci., Vol. 28 (Springer, Berlin, Heidelberg 1995) Chap. 6
[6.14] N.M. Amer, A. Skumanich, D. Ripple: Appl. Phys. Lett. **49**, 137 (1986)
[6.15] S. Grafström, J. Kowalski, R. Neumann, O. Probst, M. Wörtge: J. Vac. Sci. Technol. B **9**, 568 (1991)
[6.16] O. Probst, S. Grafström, J. Fritz, S. Dey, J. Kowalski, R. Neumann, M. Wörtge, G. zu Putliz: Appl. Phys. A **59**, 109 (1994)
[6.17] J.M.R. Weaver, L.M. Walpita, H.K. Wickramasinghe: Nature **342**, 783 (1989)
[6.18] C.C. Williams, H.K. Wickramasinghe: Nature **334**, 317 (1990)
[6.19] C.R. Leavens, G.C. Aers: Solid State Commun. **61**, 289 (1987)
[6.20] G.F.A. van de Walle, H. van Kempen, P. Wyder, P. Davidsson: Appl. Phys. Lett. **50**, 22 (1987)
[6.21] G.F.A. van de Walle, H. van Kempen, P. Wyder, P. Davidsson: Surf. Sci. **181**, 356 (1987)
[6.22] R.J. Hamers, K. Markert: J. Vac. Sci. Technol. A **8**, 3524 (1990)
[6.23] M. McEllistrem, G. Haase, D. Chen, R.J. Hamers: Phys. Rev. Lett. **70**, 2471 (1993)
[6.24] Y. Kuk, R.S. Becker, P.J. Silverman, G.P. Kochanski: Phys. Rev. Lett. **65**, 456 (1990)
[6.25] G.P. Kochanski, R.F. Bell: Surf. Sci. Lett. **273**, L435 (1992)
[6.26] D.G. Cahill, R.J. Hamers: Phys. Rev. B **44**, 1387 (1991)
[6.27] L. Arnold, W. Krieger, H. Walther: J. Vac. Sci. Technol. A **6**, 466 (1988)
[6.28] A.A. Lucas, P.H. Cutler: Solid State Commun. **13**, 361 (1973)
[6.29] C. Sammet: Aufbau und Erprobung eines Rastertunnelmikroskops im Ultrahochvakuum. Diploma Thesis, Ludwig-Maximilians-Universität München (1992)
[6.30] W. Krieger, T. Suzuki, M. Völcker, H. Walther: Phys. Rev. B **41**, 10229 (1990)
[6.31] J.-P. Thost: Einkopplung von Laserlicht in den Tunnelübergang eines Rastertunnelmikroskops. Diploma Thesis, Ludwig-Maximilians-Universität München (1989)
[6.32] T.F. Heinz: In *Nonlinear Surface Electromagnetic Phenomena*, ed. by H.-E. Ponath, G.I. Stegeman (Elsevier, Amsterdam 1991) p. 353
[6.33] C. Sammet, W. Krieger, M. Völcker, H. Walther: In *Photons and Local Probes*, ed. by O. Marti, R. Möller (Kluwer, Dordrecht 1995)
[6.34] A. Levy Yeyati, F. Flores: Phys. Rev. B **44**, 9020 (1991)
[6.35] A. Levy Yeyati, F. Flores: Ultramicrosc. **42/44**, 242 (1992)
[6.36] A. Levy Yeyati, F. Flores: J. Phys., Condes. Matter **4**, 7341 (1992)
[6.37] N.M. Miskovsky, Sookyung H. Park, P.H. Cutler, T.E. Sullivan: J. Vac. Sci. Technol. B **12**, 2148 (1994)
[6.38] R.J. Hamers, D.G. Cahill: Appl. Phys. Lett. **57**, 2031 (1990)
[6.39] S. Akari, M.Ch. Lux-Steiner, K. Glöckler, T. Schill, R. Heitkamp, B. Koslowski, K. Dransfeld: Ann. Physik **2**, 141 (1993)
[6.40] S. Akari, M.Ch. Lux-Steiner, M. Vögt, M. Stachel, K. Dransfeld: J. Vac. Sci. Technol. B **9**, 561 (1991)
[6.41] M. Völcker, W. Krieger, H. Walther: Phys. Rev. Lett. **66**, 1717 (1991)
[6.42] M. Völcker, W. Krieger, T. Suzuki, H. Walther: J. Vac. Sci. Technol. B **9**, 541 (1991)
[6.43] M. Völcker, W. Krieger, H. Walther: J. Appl. Phys. **74**, 5426 (1993)

[6.44] N. Kroo, J.-P. Thost, M. Völcker, W. Krieger, H. Walther: Europhys. Lett. **15**, 289 (1991)

[6.45] R. Möller, U. Albrecht, J. Boneberg, B. Koslowski, P. Leiderer, K. Dransfeld: J. Vac. Sci. Technol. B **9**, 506 (1991)

[6.46] C. Baur, A. Rettenberger, K. Dransfeld, P. Leiderer, B. Koslowski, R. Möller, P. Johansson: In *Photons and Local Probes*, ed. by O. Marti, R. Möller (Kluwer, Dordrecht 1995)

[6.47] A. Hornsteiner, W. Krieger, Z. Szentirmay, N. Kroo, H. Walther: In *Photons and Local Probes*, ed. by O. Marti, R. Möller (Kluwer, Dordrecht 1995)

[6.48] G.P. Kochanski: Phys. Rev. Lett. **62**, 2285 (1989)

[6.49] B. Michel, W. Mizutani, R. Schierle, A. Jarosch, W. Knop, H. Benedickter, W. Bächtold, H. Rohrer: Rev. Sci. Instrum. **63**, 4080 (1992)

[6.50] W. Seifert, E. Gerner, M. Stachel, K. Dransfeld: Ultramicrosc. **42/44**, 379 (1992)

[6.51] S.J. Stranick, L.A. Bumm, M.M. Kamna, P.S. Weiss: In *Photons and Local Probes*, ed. by O. Marti, R. Möller (Kluwer, Dordrecht 1995)

[6.52] S.J. Stranick, P.S. Weiss: J. Phys. Chem. **98**, 1762 (1994)

[6.53] A.A. Lucas, P.H. Cutler, T.E. Feuchtwang, T.T. Tsong, T.E. Sullivan, Y. Yuk, H. Nguyen, P.J. Silverman: J. Vac. Sci. Technol. A **6**, 461 (1988)

[6.54] M.J. Hagmann: J. Vac. Sci. Technol. B **12**, 3191 (1994)

[6.55] H.Q. Nguyen, P.H. Cutler, T.E. Feuchtwang, Z.-H. Huang, Y. Kuk, P.J. Silverman, A.A. Lucas, T.E. Sullivan: IEEE Trans. ED-**36**, 2671 (1989)

[6.56] R. Möller, C. Baur, B. Koslowski, K. Dransfeld: Intl. Conf. on Scanning Tunneling Microscopy, Interlaken (1991), Post-Deadline Contribution PA/1

7. Scanning Near-Field Optical Microscopy

U. C. Fischer

With 27 Figures

Light microscopy, which was invented more than 300 years ago, is a very important technique in various fields of science, especially in biology. By successive improvements of the optical components and the recent invention of the confocal microscope, imaging with a light microscope down to the fundamental diffraction limit has become possible. With a confocal light microscope a resolution of about 0.2 μm is achieved in the visible spectral range using oil immersion optics, as demonstrated in Fig. 7.1.

The interaction of light with matter is the basis of many different ways in which material specific contrast can be obtained in optical micrographs. Single molecule optical detection, spectroscopy and microscopy has recently become a very active area of research [7.1].

The implementation of sensitive optoelectronic detectors such as photomultiplier tubes, photoavalanche diodes and CCD cameras has improved the sensitivity of signal detection in a light microscope to such an extent that it is possible to detect the fluorescence of single molecules in video rate imaging. A very interesting application of light microscopy has recently been reported in this context [7.2]. The kinetics of hydrolysis of ATP (adenosine tri-phosphate) bound to a single myosin molecule was measured by recording the fluorescence

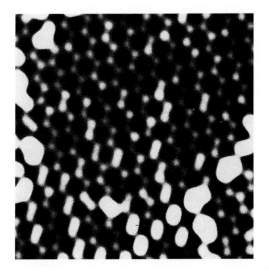

Fig. 7.1. Light micrograph of a latex projection pattern [7.184] consisting of a hexagonal array of 10 nm thick triangular aluminum patches of a center to center distance of 0.26 μm, taken with a Leica Confocal microscope at a wavelength of 514 nm using an oil immersion objective lens of a numerical aperture of 1.3. By Courtesy of U. Kubitschek, Institute of Medical Physics and Biophysics, University of Münster

of single labeled ATP molecules and evaluating the histogram of residence times of an ATP molecule on a myosine molecule. Thus, it is now possible to measure important elementary biochemical events on the level of single molecules with unprecedented sensitivity and microscopic spatial resolution by using a light microscope.

Because of the wealth of information which can be obtained by light microscopic techniques it has been a dream to break the diffraction limit in order to get optical information at a scale well below the wavelength of light.

There are schemes to extend this limit of far-field microscopy by nonlinear techniques on the basis of two photon luminescence microscopy [7.3] and 4π confocal microscopy [7.4].

There is, however, a principally different way to obtain information of optical properties on a small scale independent of the wavelength of light, by making use of the concept of the electromagnetic near-field which exists in the vicinity of any small radiating object. The concept of the near field is well known from the electric near field of a radiating electric dipole and the magnetic near field of a radiating magnetic dipole [7.5].

7.1 Background

Scanning Near-field Optical Microscopy (SNOM), using light transmitted through a small aperture in a thin metal film as a nanoscopic light source, was first proposed by *Synge* [7.6] more than sixty years ago. This light source is scanned over the surface of the sample at a distance of a few nanometers. The modulation of light emitted from the source by near-field interaction between probe and object serves as a signal for the SNOM. The resolution capability of this type of a SNOM is based on the strong confinement of light in the vicinity of the aperture. However, the proposal of *Synge* was soon forgotten.

Kuhn [7.7] suggested a non scanning imaging application of the near field in his proposal of contact imaging by energy transfer as a method for the duplication of nanostructures. Single electronically exited dye molecules act as a source of a strongly localized near field for optical information storage at nearly molecular dimension. The scheme is described in Fig. 7.2 in the form in which it was implemented experimentally [7.8]. A monomolecular film of dye molecules serves as a light sensitive film, and a very thin, only partially absorbing, planar metal pattern which is embedded into the surface of a pliable polymer film, serves as a conformal mask. The film is brought into contact with the mask and then is irradiated with light. In contrast to optical projection lithography, it does not matter from which side of the mask the film is irradiated. As a result of irradiation, the structure is transferred from the mask to the monomolecular film as a pattern of areas where the dye is bleached and where it is not bleached. After release of contact, the areas of unbleached dye correspond to the areas where there was contact to the slightly absorbing metal pattern. The resolution of this pattern transfer

Conformal Mask:
Planar metal pattern embedded into the
surface of a pliable plastic film

Lightsensitive film:
monomolecular layer of a dye

1.) Contacting

2.) Photochemical   reaction   during
contact
Result: Pattern transferred to dye layer

3) Release of contact

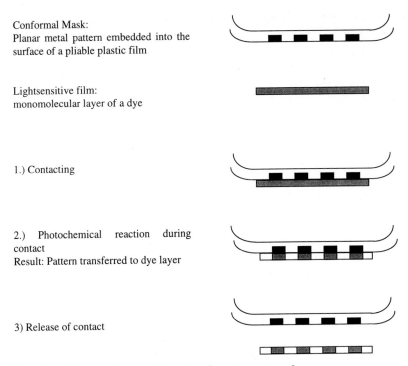

Fig. 7.2. Scheme of contact imaging by energy transfer

is not limited by the wavelength of light but by the range of the energy
transfer mechanism and the distance between the mask and the dye layer.
This process of pattern transfer is an optical imaging process which makes use
of the short range of the near field of single electronically-exited molecules in
order to obtain a resolution, which is independent of the wavelength of light. It
should, in principle, be possible to obtain molecular resolution if the sample
can be brought in molecular contact with the dye layer. Experimentally a
resolution of 70 nm was demonstrated in contact copies, which is still far
from molecular resolution.

Unaware of the original suggestion by *Synge, Ash* and *Nichols* [7.9] demonstrated the working principle of SNOM in the microwave domain. They suggested an extension to the optical spectral domain but anticipated substantial
technical difficulties.

Pohl et al. [7.10], *Fischer* [7.11, 12], and *Lewis* et al. [7.13] made an independent effort to realize SNOM in the optical spectral domain. These attempts were motivated 1) by the invention of the Scanning Tunneling Microscope by R. Binnig and H. Rohrer ([7.10]), 2) by the ideas to make use of the
near field of fluorescent molecules for imaging beyond the diffraction limit
([7.12]) and 3) by investigations of light diffraction of small microfabricated
apertures ([7.13]). Whereas *Pohl* et al. [7.10] and *Fischer* [7.12] demonstrated

submicrometer resolution in first line scan experiments, *Dürig* et al. [7.14] succeeded in obtaining SNOM images at a resolution of about 20 nm, using an improved instrument and the simultaneously recorded tunnel current between tip and sample, for controlling the distance between them during scanning.

The configuration of *Pohl* et al. [7.10] with an aperture at the apex of a transparent metal coated body has lead to the most widely used SNOM configuration. Several important improvements made possible a reproducible operation of the SNOM technique: *Haarotunian* et al. [7.15] adopted the method of thermal pulling of glass capillaries, and *Betzig* et al. [7.15] modified this method to pull quartz fibers for tip fabrication. *Betzig* et al. [7.17] and *Yang* et al. [7.18] introduced shear-force microscopy as a tool to control the distance between the tip and the object independently of the near-field optical signal. By shear-force microscopy, the topography of the sample can be obtained whereas the near-field optical signal simultaneously displays the specific optical contrast. These improvements allowed a more routinely investigation of surfaces and for the first time it was not only possible to obtain images of test objects. Mainly the group of *Betzig* at Bell Laboratories succeeded to address more relevant problems by SNOM.

SNOM with apertures is, however, only one approach to a more general concept of scanning near-field optical microscopy. In the past, several review papers on near-field microscopy have already appeared covering either the whole field [7.19–21], specific applications and developments [7.22–25] or theoretical aspects [7.26, 27]. Detailed information about the developments of near-field microscopy may also be obtained from Conference Proceedings of the International Conferences on Near-Field Microscopy [7.28–30].

This review is focused on experimental aspects of near-field microscopy with the near-field probe as the key component of a near-field microscope. An emphasis lies on recent developments indicating the potential of near-field microscopy for optical imaging at molecular resolution. An overview of applications of SNOM is given with a few selected examples, where information is obtained which cannot be obtained by other means. In some cases near-field microscopy is used not so much as a tool to obtain images beyond the fundamental diffraction-limited resolution of light microscopy, but to make use of a different contrast obtained in near-field images in comparison with far-field images. It can also circumvent technical problems which arise under certain circumstances when trying to use an objective lens of high Numerical Aperture (NA) with oil immersion.

SNOM is here the most commonly used acronym. Others in use are aperture-NSOM, PSTM (Photon Scanning Tunneling Microscope) and STOM (Scanning Tunneling Optical Microscope).

7.2 Characteristic Optical Components of a SNOM

The optical components of a SNOM are shown schematically in Fig. 7.3. In illumination modes of a SNOM, a nanoscopic tip serves as a light emitting probe and in collection modes it serves as a detector. In the illumination modes, light from a macroscopic light source is directed by ray optical components to the probe. In the case of aperture-NSOM a nanoscopic aperture in a metal film is located at the apex of the tip. The probing tip is connected to the ray optical components of the SNOM by a link which channels light efficiently from the macroscopic or microscopic dimensions of a free or guided light beam to the nanoscopic dimensions of the probing tip.The tapered part of the metal coated fiber serves as a link in the aperture-NSOM. Further ray optical components transmit light from the probe to the detector. In the collection mode, the positions of source and detector are exchanged. There are also schemes, where the optical path for illumination and detection are on the same side of the substrate. These modes are called reflection modes. The probing tip and the link are the characteristic near-field optical components of a SNOM.

Ideally, no light should leak from the ray optics of the illumination pathway to the ray optics of the detection pathway. For a very efficient SNOM, light should only be coupled into the detection pathway by the nanoscopic probing tip. However, this condition is never strictly reached, and the way in which the near-field optical components, the link, and the tip itself are made, decide how well this condition can be fulfilled. The design of the link and the tip itself are therefore the clue to a realization of an efficient SNOM. At the same time the confinement of light to nanoscopic dimension is still a challenging physical problem.

Leakage causes artefacts in SNOM images and reduces contrast but these effects can be partly overcome by modulation techniques as shown, e.g., by *Zenhausern* et al. [7.31] and *Bachelot* et al. [7.32] who did not even use a "link", but irradiated the tip directly with a focused beam of light.

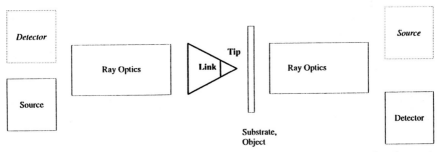

Fig. 7.3. The characteristic components of a SNOM. In collection modes the position of the detector and of the source is interchanged with respect to their position in the illumination modes

Even if the effect of leakage on the SNOM signal can be reduced, leakage itself should be reduced as much as possible to allow, e.g., optical nanolithography with an illumination mode SNOM and local photochemistry in general. For fluorescence studies, where the photodestruction limits the sensitivity of detection it is very important to avoid heating and photochemical damage by unnecessary exposure of the object. Up to now aperture-NSOM was the method with the least leakage problems. For this reason the method is suited for nanophotolithography, fluorescence studies and the measurement of locally resolved near-field induced currents.

7.2.1 Ray Optics of a SNOM

SNOM techniques differ mainly in the types of probes which are used and also by their ray optical components, i.e., their optical scheme, which is useful for a classification of most types of SNOM concepts. As shown schematically in Fig. 7.4, three different regions where the rays propagate can be distinguished: I) the body of the probe, II) the outside and III) the substrate of the object. In general, regions I and III will have a higher refractive index than the outside region II. Different angular domains of rays propagating in regions I and III exist, which can be distinguished by the criterion of total reflection of a ray falling on its boundary or of it being partially refracted into the outside II. Thus, in the case of a transparent substrate III, we distinguish between the angular domain III_1 of angles ε with $-\varepsilon_t < \varepsilon < \varepsilon_t$, where ε_t is the critical angle of total reflection ($\varepsilon_t = 41.5°$ for glass of refractive index 1.5) and the angular domain III_2 with $90° > \varepsilon > \varepsilon_t$ or $-90° < \varepsilon < -\varepsilon_t$, which is sometimes called the range of forbidden light. Rays of the domain III_2 are totally reflected in the substrate, whereas rays of domain III_1 are partially refracted into the outside II. Also within the body of the tip I two different domains may be distinguished (Fig. 7.4). This figure only shows the case of a

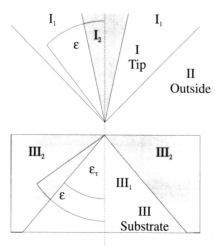

Fig. 7.4. Ray optics of a SNOM. One can distinguish between three different regions where the rays propagate; the body of the probe I, the outside II and the substrate III of the object. Different angular domains of rays propagating in regions III and I at an angle ε can be distinguished by the criterion of a ray falling on its boundary being totally reflected into the same domain (III_2, I_2) or being partially refracted into the outside II (III_1, I_1)

rectangular wedge, a two-dimensional analog of the three-dimensional body of the tip. For such a wedge, with a refractive index $n = 1.5$, rays entering at an angle within the angular domain ($-3.5° < \varepsilon < +3.5°$, region I_2) will be totally reflected back into a reflected ray of the same angle ε. Rays entering the wedge at different angles will also be reflected into the same angle and be partially refracted into the outside II of the wedge. This situation also applies, if the wedge is coated with a partially transparent metal film, as is typical for SNOM probes. Similar considerations also apply for a three-dimensional tip.

In summary, in many cases it is possible to distinguish in regions I and III between angular domains I_1 and III_1 where total reflection of the rays occurs into the same domain and the domains I_2 and III_2 from where light is partially refracted to the outside II.

Different types of SNOM have been realized, which can be classified according to the rays of the illumination pathway and those of the detection pathway propagating in different respective domains as listed in Table 7.1.

Table 7.1. Ray-optical classification of SNOM configurations according to the rays of the excitation and detection pathways propagating in the respective angular domains as defined in Fig. 7.4

Excitation	Detection	Names	Reference
I	I	Internal reflection	7.9, 67, 83–85, 130, 163, 174
I_2	I_1	Internal reflection	7.12, 43, 45, 65, 66, 69
I	II	External reflection, illumination mode	7.67, 68, 84, 107, 156, 159, 160, 162
I	III_1	Transmission, illumination mode	7.10, 14–17
I	III_2	Inverse PSTM	7.63, 64, 121, 143
II	I	External reflection, collection mode	7.162
II	II		7.32
III_1	I	Transmission, collection mode	7.188
III_1	III_1	SIAM	7.31
III_2	I	PSTM, STOM	7.60–62, 113–118
III_2	III_1		7.138
III_2	III_2	SPNM	7.87

7.2.2 The Link

As mentioned above, we consider the link between the ray optical components of a SNOM and the tip as the most characteristic component of a SNOM.

The link should channel light efficiently from the ray optical component to the tip. For the aperture-NSOM using tapered metal coated fibers as a probe, the link is the tapered part of the fiber, which connects the waveguide modes of the fiber to the tip. The tapered metal coated fiber bears analogy to a tapered cylindrical hollow metal waveguide. For ideal metals, propagating modes become evanescent and are strongly damped if the diameter of the waveguide is smaller than half the wavelength due to the cut-off condition. Although this cut-off condition is relaxed somewhat for real metals at optical frequencies, the propagation of light is attenuated over-exponentially for the section of the tip which has a diameter smaller than half the wavelength [7.33]. In order to reduce unavoidable losses in this link region of aperture probes, much effort has been spent to fabricate aperture probes with a short tapered region.

Alternative guidelines to the design of the link are derived from microwave transmission line concepts. Quite unlike the hollow waveguide, the well known coaxial cable is a device by which electromagnetic energy can be transported without significant losses through a constriction which is negligibly small compared to the wavelength. This property is due to the TEM (Transverse ElectroMagnetic) transmission line mode of the coaxial cable which has no cut-off. Moreover, the coaxial tip (a conically tapered coaxial cable) has in theory the interesting property that the electric and magnetic field amplitudes of its TEM mode increase as $1/r$ with decreasing distance r from the tip. This means that this concept allows us to increase the field intensities in the tip in comparison to the intensity by which the tip is irradiated. This model of the coaxial tip may be considered as a very effective link and tip for a SNOM [7.34–37]. The concept of the coaxial tip relies on the assumption that the metal is a perfect conductor. The validity of this concept was evaluated for real metals in the infrared domain by *Fee* et al. [7.35] and its validity in the visible part of the spectrum was discussed by *McCutchen* [7.38]. For a coaxial tip made of aluminum ending with an inner core diameter of 10 nm and an outer diameter of 50 nm, *McCutchen* estimated a $1/e$ penetration depth of 290 nm. Therefore, one can expect a validity of the concept only for a very short length of the coaxial tip. A realization of a coaxial tip for the optical domain was attempted [7.37] on the basis of metal-filled glass capillaries, the so-called *Taylor* [7.39] wires, which were thermally pulled to end in a submicrometer metal filled glass tip, but a complete fabrication scheme was not yet achieved.

A similar concept of a link is based on surface plasmons on the metal coating of a tip to confine and compress electromagnetic energy to nanoscopic dimensions in the optical spectral domain. The design of the tetrahedral tip, as described in detail below, is based on this concept.

A further concept of "funneling" light to a tip is based on the transport of electromagnetic energy by exciton migration towards a tip in "excitonic supertips" [7.40 41]. Plans to realize such tips are based on exciton conducting crystallites that absorb light and convert it to excitons, which ultimately produce photons again. Such crystallites can be grown on the tip of a tapered micropipette. An active center (ideally a single molecule) has to be positioned onto the apex of such a supertip as an efficient trap for excitons. Trapped excitons are emitted locally in the form of photons. *Kopelman* [7.28] mentioned the analogy between processes involved in the energy transfer to a tip in a SNOM probe and energy transfer processes in the photosynthetic apparatus of plants and photosynthetic bacteria where the light harvesting antenna transfer excitation energy efficiently to the molecular reaction centers [7.42].

7.2.3 The Tip

The nanoscopic tip itself is the second characteristic component of the SNOM. It serves as an antenna of nanoscopic dimensions which interacts with the object. The radiation from this antenna is used as a signal for SNOM. The near-field distribution of this antenna determines its radiative signal. As the intensity of the radiation of such an antenna scales with the 6th power of its dimensions, this radiative signal will become very small for nanoscopic tips and might well be limiting for a high resolution with a SNOM due to a very small signal to noise ratio.

An increase in scattering efficiency of an antenna can be achieved by resonance excitation of the antenna, e.g., the excitation of a local plasmon of a metal particle acting as an antenna [7.43–45]. Resonance excitation, however, reduces the spectral bandwidth in which the antenna can be used compared to a non resonantly excited antenna.

Alternatively the antenna may be coupled to a resonator [7.9, 46] or to a laser [7.47] to increase its efficiency.

Zenhausern et al. [7.48] point out, that the r^6 dependence of the radiation efficiency can be reduced by interference techniques, where, due to the measurement of the electric field amplitude rather than its intensity, the signal scales with the third power of the dimension of the probing tip.

A very interesting idea for obtaining a better signal was put forward by *Fee* et al. [7.35] in connection with the transmission line concept of a link. Whereas only a minute fraction of the light coming to the tip from the transmission line is scattered from the tip, most of the light will be reflected. The amplitude and phase of the reflected light will be strongly influenced by the local dielectric environment of the tip. The measurement of the amplitude or phase of the reflected light in internal reflection modes of a SNOM may therefore lead to very favorable SNOM signals. These produce a high resolution in combination with a good contrast, if the realization of SNOM probes based on transmission line or surface plasmon concepts of a link is successful.

The interaction of the antenna with a dielectric object is determined mainly by the electric component of the near field. The near field of a radiating source may have a predominant electric or magnetic component as is well known from the model of a dipole. An electric dipole has a strong electric near field whereas the magnetic near field is small. On the other hand, a magnetic dipole has a strong magnetic near-field and a weak electric near-field. For example, it is important that an electric dipole type of excitation, which is associated with a high local electric field rather than a magnetic dipole type of excitation, is generated on an antenna [7.9]. In the case of aperture-NSOM, the aperture serves as the antenna. The dipole nature of the excitation of an aperture in a planar metal film depends strongly on the way in which the aperture is excited. The radiation field of an aperture excited by s-polarized light (polarized perpendicular to the plane of incidence) corresponds, according to the theory by *Bethe* [7.49], to that of a magnetic dipole, whereas the radiation field of an aperture excited by p-polarized light (polarized within the plane of incidence) corresponds to the superposition of a magnetic and an electric dipole. In their microwave experiments, *Ash* and *Nichols* [7.9] observed considerable differences in image contrast depending on the way in which the aperture was excited. So far, not much effort has been spent to investigate experimentally the magnetic and electric components of the near field of SNOM probes in the optical domain. Experiments with sub-micrometer apertures as well as with metallic protrusions in a planar metal film acting as emitting near-field probes were made. Strong differences in the response of the SNOM signal of such probes interacting with dielectric material were indeed observed for p-polarized and s-polarized excitations, respectively [7.12, 45]. Experiments concerning the electric near-field distribution of an aperture probe by *Betzig* and *Chichester* [7.50] and other experiments concerning the electric and magnetic dipole nature of its radiation field by *Obermüller* and *Karrai* [7.51] are described below.

The resolution which can be obtained in SNOM images depends strongly on the effective size of the emitting or detecting probe. An aperture cannot be made much smaller than 50 nm and the spread of the near field of an aperture is determined not only by its size but also by the skin depth of the surrounding metal. It is therefore essential to use a more simple structure such as a purely dielectric or metallic tip in order to obtain a resolution significantly below 50 nm.

7.3 Distance Control in SNOM

During a scan the probe has to be kept at a close distance to the sample in a controlled fashion, because the distance between probe and sample limits the obtainable resolution and crashes of the tip with the sample may destroy the probe. First, simultaneous STM modes, requiring a conductive probe and sample were and still are used occasionally [7.14, 52]. Combining SNOM with an AFM (Atomic Force Microscope) mode using the same tip turned out to be the most versatile way to control the distance. AFM is usually performed with tips on a microfabricated cantilever vibrating perpendicular to the sample. *Betzig* et al. [7.17] and independently *Yang* et al. [7.18], succeeded to combine aperture-NSOM with tapered fiber probes and AFM by introducing shear-force microscopy where the fiber oscillates parallel to the surface. As the tip approaches the surface of the sample the oscillation of the tip is influenced by interactions of the tip with the sample. These interactions lead to a change in the amplitude and phase of vibration which are detected by optical or alternatively by piezoelectric detection schemes [7.53]. The interaction has a range of 1–100 nm and is well suited to control the distance between tip and sample. At the same time a topographic image of the surface can be obtained simultaneously with the optical signal. It is thus possible to correlate optical and topographic information, which is considered to be an important advantage of near-field microscopy. The interaction responsible for the shear force signal is not fully understood, but a number of possibilities have been put forward, including long-range van der Waals forces, image charge current dissipation, viscous damping in a contaminant layer and short range mechanical contact (knocking) [7.54–56]. The chemical composition of the surface of the sample seems to play an important role for the shear force signal too [7.57]. The introduction of shear force microscopy as a means to control the distance was the trigger for a fast development of aperture-NSOM into a widely used method, which is now also available in commercial instruments.

There are, however, other approaches to control the distance in SNOM by an auxiliary AFM mode. *Lieberman* et al. [7.58] and Ataka et al. [7.59] introduced aperture probes based on bent tapered fibers for contact mode AFM. The bend of the fiber is flattened and polished and can thus serve as a mirror for force detection by beam deflection.

A further approach can be seen in the integration of a SNOM probe into a microfabricated AFM cantilever (see below).

In several cases SNOM is performed in a purely optical mode, where the distance is also controlled by the SNOM signal. Thus, according to *Courjon* et al. [7.60] and *Reddick* et al. [7.61], in the STOM (Scanning Tunneling Optical Microscope) or PSTM (Photon Scanning Tunneling Microscope) configuration, the exponential decay of the evanescent modes of a dielectric interface are used to obtain an optical signal which strongly depends on the distance between sample and object. A scheme of the PSTM mode is shown in Fig. 7.5. The PSTM mode is a collection mode SNOM where the probe serves as a local detector of light. Evanescent modes, generated by total internal reflection, are used to irradiate the surface of the sample. A tip at the end of a tapered glass fiber diving into the evanescent field is excited by this field and couples light into propagating modes of the fiber which are detected as a signal. The signal increases exponentially with the tip approaching towards the surface. The slope depends on the wavelength of the exciting light, the reflection angle of the exciting light and the refractive index. In a constant intensity mode the signal is used to keep the probe at a controlled distance to the surface of the sample similar to STM. The deviation of the probe from the mean distance is used as a signal. In a first approximation the topography of the sample is

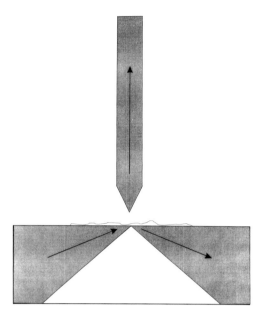

Fig. 7.5. Scheme of PSTM. The sample is illuminated by total internal reflection.When the detecting tip penetrates into the range of the evanescent modes, light is coupled into the fiber by the tip

retrieved by this signal. The optical signal depends however not only on the topography but also on the local refractive index of the sample and therefore is no reliable measure of the distance between tip and sample. A separation of optical and topographic components is experimentally achieved by a distance modulation technique [7.62] where the dc signal is used to obtain information about the topography of the sample, whereas the modulated signal contains information about the refractive index.

The inverse PSTM [7.37, 63, 64] is an illumination mode SNOM where the tip is used as local emitter rather than a detector as shown in Fig. 7.6. As the tip approaches the sample, the tip excites evanescent modes which radiate into the angular domain of total internal reflection (forbidden light). In the inverse PSTM, only the forbidden light is used as a signal, which is collected by a dark field immersion collector. Such a collector may consist of a combination of a hemisphere and an elliptic mirror [7.63],as indicated in Fig. 7.6, or by, for example, a parabolic mirror filled with dielectric material [7.64]. The signal increases exponentially with decreasing distance between probe and sample. It was demonstrated, using the tetrahedral tip as a probe that this signal can be used to control the distance during a scan [7.64].

A more pronounced dependence of the optical signal on distance between probe and sample was generally observed in reflection mode SNOM configurations [7.12, 65–69] in comparison to transmission mode SNOM configurations [7.19]. This dependence, which seems to be related to local evanescent fields of the probe, can also be exploited to control the distance in a purely optical mode of operation of a SNOM [7.66, 69].

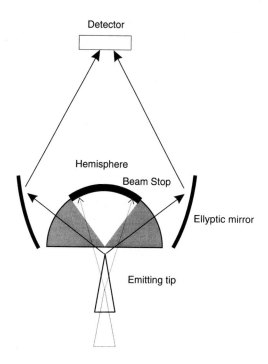

Fig. 7.6. Scheme of inverse PSTM [7.63]. The interface of a hemisphere serves as a substrate. The light, which is emitted into the hemisphere into the angular range of total reflection (forbidden range) is deflected by an elliptic mirror to the detector. The light entering the hemisphere at other angles is absorbed by a beam stop. When the tip penetrates the range of the evanescent modes of the interface, light is emitted from the tip into the forbidden range

In all these cases, where no auxiliary signal is used for distance control, the SNOM signal does not only depend on the distance between tip and sample, but also on the local optical properties and topographic features of the sample which often cannot be separated; therefore these methods can only be applied in few selected cases.

7.4 Artifacts

In SNOM imaging auxiliary AFM or STM signals are often used for an independent control of distance during scanning, as described above. These auxiliary methods have a very high lateral resolution. Only apparently high resolution can be obtained in SNOM images due to a crosstalk between a highly resolving AFM or STM imaging mode with a SNOM signal of much coarser resolution. *Hecht* et al. [7.70] worked out criteria for distinguishing between apparent and true resolution in SNOM images. One way to obtain a conservative estimate of the SNOM resolution is to perform scans at a constant height without operation of the feedback control. The auxiliary feedback is only used to approach the tip to the sample and to evaluate an operating height of the tip during a scan where no crash occurs.

By a scan without feedback control, one can obtain an upper limit for the resolution in SNOM imaging. On the other hand the full resolution potential of SNOM can only be obtained employing an independent feedback in the case of not ideally flat samples. Here, criteria have to be found in order to discriminate between true and only apparent resolution in SNOM images. Such criteria may be based on an analysis of the SNOM image $S(x, y, z)$ and of the simultaneously recorded topographic image $z(x, y)$ of the sample on one hand and the dependence of the SNOM signal as a function of distance d between probe and sample $\delta S/\delta d$ on the other. If $S(x1, y1, z1) - S(x2, y2, z2)$ strongly differs from $(\delta S/\delta d)(z1 - z2)$ artifacts can be excluded. If there is no difference, the variation of the SNOM signal may well be due to a non local dependence of the optical signal on distance between probe and sample, which cannot be considered as a true SNOM signal.

7.5 Probe Concepts

The development of SNOM probes and their efficient fabrication are certainly the most important steps for further progress in near-field microscopy. One may distinguish between passive and active probes. In the active probe a function such as an optoelectronic detector is integrated into the probe thus leading to an important simplification of the SNOM scheme of Fig. 7.3. In this case the detection pathway can be omitted completely. Microfabrication techniques are desirable for obtaining cheap and reproducible probes. In this chapter various schemes of passive probes, active probes and microfabricated probes are discussed.

7.5.1 Passive Probes

Various types of passive probes have been developed. SEM micrographs of four examples of such probes are shown in Fig. 7.7 as examples for different types which will be discussed in more detail below: a) an aperture probe made on the basis of a thermally pulled quartz fiber, b) a chemically etched quartz tip, c) an aperture probe based on a chemically etched quartz fiber and d) the tetrahedral tip.

7.5.1.1 Aperture Probes. Aperture probes are the most frequently used SNOM probes. For their fabrication, *Harootunian* et al. [7.15] adopted the method of pulling capillaries to fine tips, and *Betzig* et al. [7.16] modified it to pull tapered optical quartz fibers ending in a small tip with a flat end of a diameter of 20–200 nm. The taper angle and the diameter of the flat end of the tip can be determined by the pulling parameters. The fiber is then coated at an oblique angle with a layer of aluminum by thermal evaporation, such that the end face of the tip remains uncoated and thus forms an aperture. If light is coupled into the end of the fiber, a small fraction of this light is emitted from the confined area of the aperture. In order that the light is well confined to the dimensions of the aperture, the metal coating has to be quite thick [7.33] and is usually chosen to be in the order of 100 nm. A detailed

(a)
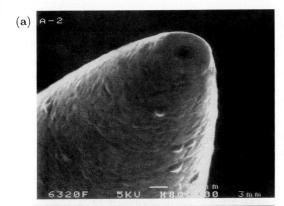

(b)

(c)
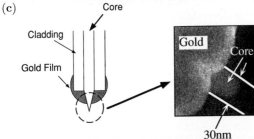

Core

Cladding

Gold Film

Gold

Core

30nm

(d)
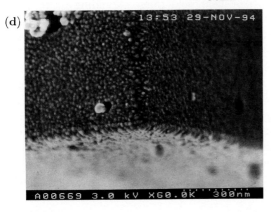

Fig. 7.7a–d. SEM micrographs of SNOM probes: (**a**) Tip of an aperture probe consisting of a thermally pulled tip of a quartz monomode fiber coated with aluminum. (By Courtesy of Sunney Xie). (**b**) Etched fiber tip according to Ohtsu [7.23]. The tip is fabricated by wet etching of a monomode quartz fiber. The thickness of the fiber coating is strongly reduced at the end of the fiber and a sharply pointed tip sticking out from the end is formed from the core. (**c**) Aperture probe fabricated on the basis of an etched tip, as shown in (**b**). The etched tip is coated with gold which is removed from the apex of the tip by a lithographic process such that a small aperture is formed with the tip sticking out [7.117]. (**d**) Tetrahedral tip. The tetrahedral tip consists of a glass fragment which is coated with metal. By courtesy of, R. Reichelt, Institute of medical Physics and Biophysics, University of Münster

description of the parameters influencing the fabrication of aperture probes was given by *Valaskovic* et al. [7.71]. Although the dimension of the aperture may be as small as 20 nm, the overall diameter of the tip is more than 200 nm. The distribution of the intensity of the electric field in the vicinity of such an aperture probe of a diameter of 100 nm was determined in a series of experiments by *Betzig* and *Chichester* [7.50]. They obtained the first convincing microscopic images of single molecules with their aperture-NSOM. It turned out that the images of single molecules reflect the distribution of the near field in the vicinity of the aperture probe. The electric field polarized within the aperture plane is rather uniformly distributed, whereas there also exists a component polarized perpendicular to this plane at the rim of the aperture. The far-field radiation of such an aperture was investigated by their angular radiation pattern which is a reproducible indicator of the quality and the size of the apertures down to 60 nm [7.72]. The dipole nature of this far-field radiation pattern corresponds to the superposition of an electric dipole and a magnetic dipole, twice as large as the electric one, both being oriented within the plane of the aperture parallel to the orientation of the transverse components of the exciting light beam. The dipole nature of the far-field radiation is different from the one predicted by the theory of scattering from small holes in a planar infinitely thin sheet of a perfect conductor [7.49] which corresponds to a magnetic dipole for normal incidence of the exiting radiation on the sheet, whereas an electric dipole is induced only for exiting light at oblique incidence by the component of the electric field polarized perpendicular to the sheet. The induced electric dipole is then oriented perpendicular to the plane of the aperture. Experimentally, the polarization of the far field of the light emitted from the aperture corresponds to about 97 % to the polarization of the exciting light [7.51]. The intensity of light transmitted to the far field ranges from 1 nW for an aperture of a diameter of 60 nm to 100 nW for 200 nm when 1 mW of 633 nm light is launched into the fiber. *Obermüller* et al. [7.72] and *Karrai* et al. [7.73] measured also the signal detected by a photodiode as a function of distance between a light emitting aperture and the front face of the detector for an aperture diameter of 90 nm (Fig. 7.8). The detected signal strongly depends on distance. They measured a 7.7 fold increase of the detected signal in the near field range extending to about 100 nm. *Obermüller* et al. [7.72] interpreted their data as an indication of the short range of the near field of an aperture. From their data they estimate the enhancement by a factor of 90 of the near-field intensity compared to the intensity of the far-field component. They estimate a decay length of 20 nm of the near field. It is however, not evident that these data of the photodetector as a function of distance between probe and surface of the detector can be interpreted as a characteristic property of the near field of the probe only, because the detector function is not spatially limited to the surface of the detector but extends to a fraction of a micron into the bulk. A pure interface effect may therefore contribute to the sharp rise in detected intensity as the probe approaches the surface similar to the exponential rise of the signal

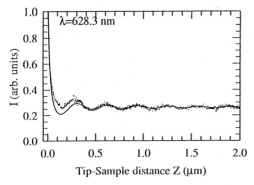

Fig. 7.8. Photocurrent measured as a function of tip sample separation Z for an aperture probe similar as shown in Fig. 7.7a with an aperture diameter of 60–90 nm. The exponentially decreasing signal is due to the aperture near-field intensity contribution to the photocurrent. In the far-field, small oscillations with a period $\lambda/2$ ride over a constant photo-signal background. The *full line* is the result of an elementary model calculation in which the subwavelength aperture is approximated by a magnetic point dipole oriented perpendicular to the tip axis. The oscillations are the interference pattern between such a dipole and its image in the detector [7.73]

in the inverse PSTM configuration discussed above [7.63, 64]. Light emitted from the probe starts to be emitted into the forbidden angles (Region III$_2$ of Fig. 7.4) when the probe penetrates into the range of evanescent modes of the interface, which for a semiconductor of high refractive index are expected to have a very short range.

In such an aperture probe the tapered part of the metal coated fiber has the function of the link between the fiber and the aperture. Since the light is over-exponentially attenuated in this link [7.33], the taper of the fiber should be as short as possible. With chemical etching as an alternative way of tip fabrication, one has more freedom to produce tips with a short taper. The method was elaborated extensively by *Ohtsu* and coworkers and their methods of tip fabrication are reviewed in some detail elsewhere [7.23, 74–77]. In their case, the aperture is made by coating the entire tip first with metal, aluminum or gold, and then removing metal at the tip with either a lithographic or a photo-lithographic technique [7.23]. In their probes a small quartz tip sticks out of the aperture. They succeeded to vary conditions of fabrication either to obtain aperture probes showing a high lateral resolution or rather a high transmission at the expense of resolution [7.77]. Also *Zeisel* et al. [7.78] succeeded to produce aperture probes on the basis of chemically etched fiber tips with a strongly increased transmission compared to the ones on the basis of thermally pulled fiber tips. The high transmission is especially important for collection mode imaging with a SNOM, fluorescence imaging and for light induced modifications with a SNOM.

Aperture probes are well suited for fluorescence studies and for nanophotolithography due to their low leakage. On the other hand, it is difficult to
obtain a resolution below 50 nm with such probes and the resolution seems
to be limited to about 20 nm.

Apertures smaller than 20 nm are very difficult to fabricate and it is doubtful whether a better resolution can be obtained with such probes, because the
penetration of light through an aperture is also determined by the skindepth
in the order of 10 nm. *Novotny* et al. [7.33] predict that a confinement might
be improved by a design where the aperture is partly covered with metal. No
systematic experimental investigation has been performed so far.

7.5.1.2 The Tetrahedral Tip. The concept of transmission lines without
cut-off for the design of a probe of SNOM with the property of an effective
link for confining electromagnetic energy from macroscopic to nanoscopic dimensions is not limited to the coaxial tip (Sect. 7.2.2). In this section we
provide plausible arguments of how the tetrahedral tip [7.64] may serve as
a photonic device for the focussing of light to nanoscopic dimensions. An
idealized scheme of a tetrahedral tip is shown in Fig. 7.9. The body consists
of a glass prism with three faces and edges converging to a common corner
which serves as a nanoscopic light emitting tip. The faces are coated with
a thin film of metal, preferentially gold, silver or aluminum and one of the
edges (K_1) is assumed not to be coated with metal. The tip is also coated
with metal. Such a tetrahedral tip contains several elements which may support electromagnetic surface waves which allow us to confine and concentrate
electromagnetic energy in three steps by "compression" in the transition

1) of an incoming 3-dimensional beam to a 2-dimensional surface wave on
   metallized faces S_{12} and S_{13} of the tip

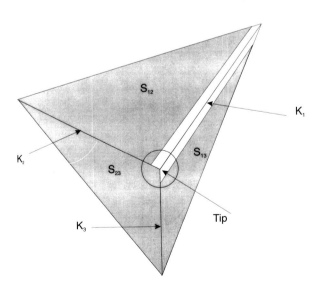

Fig. 7.9. Scheme of
a tetrahedral tip. A
corner of a glass frag-
ment forms the body
of the tetrahedral tip.
It is coated with
metal by two succes-
sive evaporation steps
such that the edge
K_1 between two frac-
ture faces S_{12} and
S_{13} is expected not to
be coated with metal.
The two other edges
K_2 and K_3, the face
S_{23} and the tip are
coated with metal

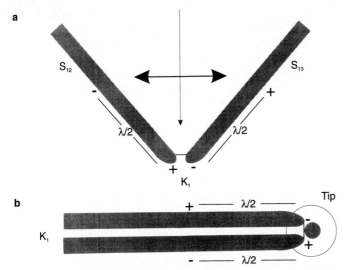

Fig. 7.10a, b. Model of light compression in the tetrahedral tip: (a) A beam of light incident within the tip excites surface plasmons on the faces S_{12} and S_{13}. (b) The surface plasmons propagate towards the edge K_1, where they excite an antisymmetric edge mode which propagates towards the tip

2) of these 2-dimensional surface waves to a 1-dimensional "transmission line" mode – a 1-dimensional surface plasmon along the edge K_1 – and
3) of this 1-dimensional mode to a localized excitation of a metal particle at the tip [7.25].

In the first step, the incoming beam can excite surface plasmons on the metal films on the faces S_{12} and S_{13} by a Kretschmann type of configuration [7.79] (Fig. 7.10a). These surface waves are localized on the outside of the metal films on the faces of the prism and they leak out to a distance of a fraction of the wavelength. Figure 7.10 illustrates how the surface waves are excited by an incoming beam which is directed obliquely with respect to the edge K_1, travel towards this edge and may excite a linear wave there traveling along the gap between the metal films. This gap should have similar properties as a parallel wire transmission line as described for ideal (and to some extent also for non ideal metal wires) by *Sommerfeld* [7.80]. Such parallel wire transmission line modes are strongly localized to the gap between the "wires", with their localization not being determined by the wavelength of this traveling wave – as is the case for the two dimensional surface wave – but rather by the width of the gap between the "wires" of a few tens of nanometers. Figure 7.10b illustrates the case of a mode consisting of two polarization waves traveling out of phase along the two "wires". In the third step this mode excites a localized mode on a metal grain at the end of the "double wire" line, leading to a strongly localized near-field. The far field of this localized mode is detected as a signal for SNOM.

The assumption of strongly localized linear surface plasmon modes along the edge of the two converging metal films is plausible because it is known that phonons can exist on a similar structure, the edge of a solid dielectric wedge [7.81, 82]. The model of light confinement and compression by the tetrahedral tip is only a very qualitative one. It was, however, a useful guideline for the fabrication of a SNOM probe which was implemented first in an inverse PSTM mode [7.64], a transmission mode [7.52] and in the meantime also in an internal reflection mode [7.83].

For a realization of the tetrahedral tip [7.64], a triangular glass fragment is formed from a microscope cover glass by breaking the glass twice such that two fracture planes S_{12} and S_{13} form a common nm sharp fracture edge K_1. Together with two other sharp fracture edges K_2 and K_3 including a segment S_{23} of the flat surface of the cover glass, the edge K_1 yields into a nm sharp corner which forms the body of the tetrahedral tip. This corner has the form of the tip of a tetrahedron and is therefore called the tetrahedral tip. The whole tip is coated by a two stage metal evaporation process, such that only the edge K_1 is uncoated or coated with less metal than the rest of the tetrahedral tip. Figure 7.7d shows a high resolution SEM micrograph of a tetrahedral tip as used in SNOM experiments with gold as a metal.

7.5.1.3 Opaque Tips. Several schemes of collection mode SNOM do not use a specially designed tip but simply a transparent quartz tip, or opaque tips such as an AFM tip or a metallic tip as a probe. As discussed above, the transparent tips are mainly used in the PSTM mode (Sect. 7.3). But these transparent tips are occasionally used in internal reflection mode configurations, too [7.84, 85]. *Fischer* [7.86] suggested to use opaque tips as used for STM in a configuration, where the tip is irradiated as in the PSTM configuration by the evanescent modes due to total internal rerflection at the sample-air interface. The light scattered from the tip and transmitted through the sample is collected as a SNOM signal. Such concepts were realized using other schemes of illumination, AFM tips, and refined interferometric modulation schemes for signal detection by *Zenhausern* et al. [7,31], *Bachelot* et al. [7.32] and by *Specht* et al. [7.87]; and this is described below (Sect. 7.6.2.4).

7.5.1.4 Microfabricated Probes. Some effort is put into the microfabrication of SNOM tips by procedures allowing mass production. The motivation of such efforts is twofold. A success of such a scheme would be an important step in making SNOM a routine technique for the submicroscopic optical surface characterization. Furthermore, the miniaturization of a SNOM probe and its integration into a cantilever-type AFM probe would make the SNOM compatible with very well developed AFM instrumentation. A microfabrication scheme for hollow pyramidal tips with an aperture at the apex was first developed by *Prater* et al. [7.88] in connection with sensors for the scanning ion conductance microscope. *Münster* et al. [7.89] arrived at the fabrication of a metallic pyramidal hollow tip on a cantilever with a rectangular aperture at the tip. Such a tip is shown in Fig. 7.11. It was already possible to perform

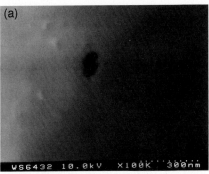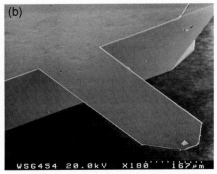

Fig. 7.11a, b. Microfabricated SNOM/AFM probe on the basis of a hollow metallic pyramid consisting of chrome with an aperture at the apex of the pyramid, which is attached to a silicon cantilever [7.89]. **(a)** Aperture at the tip of a pyramid. The apparent oblong shape is due to a drift. The true shape should be circular. **(b)** silicon cantilever with pyramid

simultaneous SNOM and AFM imaging at a resolution of about 100 nm with such tips. Alternatives are microfabricated aperture probes on the basis of conical Si_3N_4 tips. *Ruiter* et al. [7.90] introduced a scheme of microfabricating such tips on a Si_3N_4 cantilever after first attempts to use commercial AFM Si_3N_4 cantilevers with pyramidal tips directly as SNOM probes [7.91]. *Noell* et al. [7.92] produced conical Si_3N_4 tips which are coated with a thin film of aluminum, leaving an aperture at the tip with a Si_3N_4 tip sticking out of the aperture similar as in the case of the aperture probe of *Ohtsu* [7.23]. Such a tip is shown in Fig. 7.12. Again, the aim of their fabrication scheme is an integration of a tip into an AFM cantilever, but in addition they intend to integrate a strip-waveguide structure into the cantilever and a link to a fiber for a simple to use connection to the end of a cleaved fiber. The fabrication scheme of this integrated SNOM sensor is not yet completed. The microfabricated conical Si_3N_4 aperture probes have the desirable wide

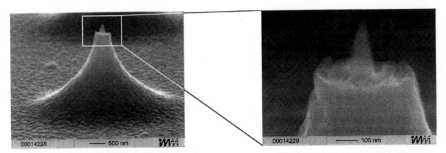

Fig. 7.12. Micromachined aperture probe tip for AFM/SNOM: The aperture is formed by the opaque Al coating around the smooth surface of the protruding and transparent Si_3N_4 tip core. The aperture diameter of the tip depicted in the left SEM image is 120 nm [7.92]

opening angle with a very short cut-off region such that they are expected to have a very high transmission compared to fiber-based aperture probes with a more extended taper. In addition, Si_3N_4 has a high refractive index of 2.4 and therefore cutoff appears for a vacuum wavelength of 670 nm only at a diameter of 140 nm. Hence, the choice of this material is also very favorable for obtaining a high transmission of such probes.

7.5.2 Active Probes

With their scanning thermal profiler *Williams* and *Wickramasinghe* [7.93] first realized a tip where a signal transducer – a thermocouple in this case – was directly integrated into a probing tip. The thermoprofiler may be considered as a near-field microscope in the frequency region of thermal radiation.

Also in SNOM it is an intriguing idea to integrate the function of a detector or a light source into the tip. Considering that a luminescent molecule constitutes a nanoscopoic light source, the idea was put forward to integrate a photoluminescent center consisting of an aggregate of molecules [7.94] or an inorganic luminescent center such as particle of porous silicon [7.95] or a color center in a LiF crystal [7.96] into a probing tip. If the luminescent centers are excited by light, the emitted light can be separated spectrally from the exciting light [7.97]. An interesting aspect of such luminescent centers is their function as a sensor for other than optical properties of their near-field environment. By using a pH indicator as a luminescent center for example, it is possible to measure the pH in an extremely small volume [7.98]. A yet more challenging aim is the integration of an electroluminescent center into a tip. There has been some effort in this direction on the basis of a complex fabrication scheme using a luminescent ZnS/MnCu powder by *Kuck* et al. [7.99]. A voltage of the order of 200 V was applied in order to observe a rather weak luminescence of a microscopic spot. A light emitting nanoscopic edge – not yet a tip – was realized by *Danzebrink* and *Fischer* [7.100], presumably on the basis of an avalanche luminescence of Schottky barriers starting at a voltage of about 5 V. The electroluminescence originates from an uncoated nanoscopic slit at the edge between two adjacent metal coated cleavage faces of a silicon or GaAs wafer. No serious attempt has been made to use this rather weak and unstable emission as a source for near-field microscopy.

The same cleaved semiconductor device could be used as a nanoscopic slit detector [7.101], and its reliable function makes clear that the integration of an optoelectronic detector into the SNOM tip is a more promising approach than the integration of a light emitting electroluminescent tip. *Danzebrink* et al. [7.102], *Danzebrink* [7.103], *Davis* et al. [7.104], *Davis* and *Williams* [7.105], and *Akamine* et al. [7.106] made an effort to integrate an optoelectronic SNOM detector directly into an AFM tip. In the schemes of *Williams* and co-workers and of *Danzebrink* and co-workers, as shown schematically in Fig. 7.13., the sensor consists of silicon tip coated with aluminum with an

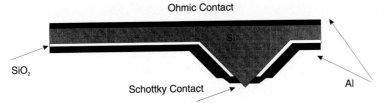

Fig. 7.13. Scheme of microfabricated active SNOM probe with integrated detector function on the basis of a Schottky diode

aperture in the metal film at the apex of the tip. The metal-semiconductor interface constitutes the light sensitive Schottky diode. *Wiliams* and co-workers introduced an SiO_2 buffer layer in order to reduce the size of the Schottky diode and in order to reduce background noise. The detector thus consists of a Schottky barrier with a small hole. Only the light which penetrates the hole contributes to the signal. The second electrode is remote from the first one. The Schottky detector may be considered a detector of the near-field intensity of the aperture. *Williams* produced the hole by a lithographic technique similar to the one employed for the passive aperture probe by *Ohtsu* [7.23]. The tip is not yet incorporated into a cantilever. First scanned images were obtained with such a tip with a resolution in the order of 100 nm using a PSTM set up with simultaneous contact force imaging of a sample attached on a cantilever as a force sensor. *Danzebrink* [7.103] modified a tip on a Si cantilever as standardly used for AFM, coated it with Al and then produced an aperture by focussed ion beam lithography. It seems very likely that the integration of a Schottky barrier will lead to a complete scheme of a SNOM probe with an integrated optoelectronic function.

Akamine et al. [7.106] pursued a somewhat different concept. The detecting function of a pn-diode is not directly located at the tip but rather at the flat face of the cantilever. The function of this detector is therefore to pick up the scattered field from the tip.

7.6 Contrast in Near-Field Microscopy

With SNOM, optical properties can be determined at a lateral resolution well below the diffraction limit of classical light microscopy. In some cases information about optical properties can be obtained by the same contrast as in far-field microscopy, which we call conventional contrast. In this case the advantage of a near-field microscope will be in imaging of optical properties at a smaller scale than the conventional light microscope. Due to the better localization, it is possible to observe optical properties of individual sites which are obscured in the observation of large population due to an overlap of the properties of individuals. In near-field microscopy the proximity of the probe to the sample leads however to alternative ways of obtaining optical contrast which depend on an interaction of the probe with the sample. Such a contrast will be called specific near-field contrast.

7.6.1 Contrast in the Conventional Sense

Many different types of contrast have been demonstrated in SNOM which are known from conventional light microscopy. They are summarized in a review by *Trautman* et al. [7.107] and by *Moers* et al. [7.108] among them there are absorption contrast, phase contrast, polarization contrast, fluorescence and magnetooptical contrast. Even Raman spectroscopy [7.109–111] with a SNOM was attempted.

7.6.2 Specific Near-Field Contrast

In conventional microscopy the imaging properties are described by a transfer function of the object and by an independent point spread function of the microscope. The transfer function depends on the object and the way it is illuminated and the point spread function is independent of the object. Attempts were made to describe the SNOM imaging process by a similar approach [7.112]. On the other hand, the near-field interaction between probe and sample is the prerequisite to imaging beyond the diffraction limit by SNOM. Specific near-field contrast is discussed in the following paragraphs.

7.6.2.1 Evanescent Fields at Extended Interfaces. Evanescent fields exist at extended interfaces between two media of different refractive index. They can only be detected indirectly by far-field methods due to scattering of small particles immersed in the evanescent fields. A near-field probe may replace such a scattering particle and thus the evanescent field becomes visible in a SNOM micrograph. The configuration of the PSTM described above is based on this concept. In this way the evanescent field induced by total reflection at interfaces or at waveguide structures [7.113] or of light emitting devices can be measured and the resonant field enhancement of the evanescent surface plasmon modes can be determined [7.114] .

7.6.2.2 Local Evanescent Fields. The near field of a microscopic particle is strongly localized to its surrounding. Also this near field can be made visible if a scattering object, e.g., a SNOM probe is placed as a detector into its near-field range. On the other hand, an emitting SNOM probe as used in an illumination mode configuration has a strongly confined near-field in its surrounding. This near field is manifested in the influence of a scattering object within the near-field range of the probe on the SNOM signal.

As the local evanescent field components can only be made visible by perturbation, the SNOM signal reflects the properties of both, the probe and the sample. It is therefore not a simple task to extract the desired information about a sample.

7.6.2.3 Local Evanescent Field of the Sample. In a PSTM configuration using pointed glass fibers without metal coating as a probe, *Weeber* et al. [7.115] investigated the near-field of individual rectangular posts on a glass surface made by microfabrication techniques. In the PSTM images (Fig. 7.14a, b) they observed strong differences in the confinement of the near field for different polarizations of the incident beam. Figure 14c shows in comparison an AFM image of the posts. Only for the case of p-polarized incidence with a large component of the electric field oriented perpendicularly to the interface, a strong confinement of the near field to the dimensions of the object was observed. There is an inversion of contrast for s-polarized illumination and the shape of the near field strongly differs from the topographical shape. Line scans across the posts as shown in Fig. 7.14d clearly reveal this difference. The experimental results are explained with theoretical models of the near-field distribution around the posts. Theory predicts that the confinement becomes better when the structures are smaller. This is also observed experimentally (not shown here). The PSTM method seems to be well suited to investigate very small structures in a dilute distribution on a surface. Larger structures and a crowding of structures on a surface leads to multiple scattering phenomena, rendering the interpretation of PSTM images very difficult in these cases.

The local distribution of the near field of other small objects was also investigated. *Saiki* et al. [7.116] measured the spread of the near field of small spherical fluorescent particles in a PSTM configuration. *Naya* et al. [7.117] investigated the near field of flagellar filaments of 25 nm diameter adsorbed to a glass surface in a PSTM configuration, using a metallic probing tip (Fig. 7.7c). Also in this case the contrast is different for p- and s-polarized illumination respectively. But the contrast is inverted compared to the contrast in the images of *Weeber* et al. [7.115] described above.

Several experiments were devoted to the investigation of local perturbations of surface plasmons on a thin metal film and to optical properties of local surface plasmons [7.118–124].

7.6.2.4 Local Evanescent Field of the Probe and Probe Sample Interaction. *Betzig* and *Chichester* [7.50] recorded the near field of an aperture probe using the fluorescence of single molecules as a probe for the near field as already mentioned above. Not in all cases the fluorescence of a single molecule can be regarded as an indicator of the local field intensity because the fluorescence properties of the molecule are strongly influenced by the vicinity of the probe due to energy transfer processes. It is thus possible to investigate the interaction of a molecule with its environment on the basis of single molecule detection. Such effects were previously investigated only on large ensembles of molecules [7.125]. *Trautman* and *Macklin* [7.126] and *Bian* et al. [7.127] were able to demonstrate the influence of energy transfer processes from a single molecule to a probing tip with a near-field microscope by measuring

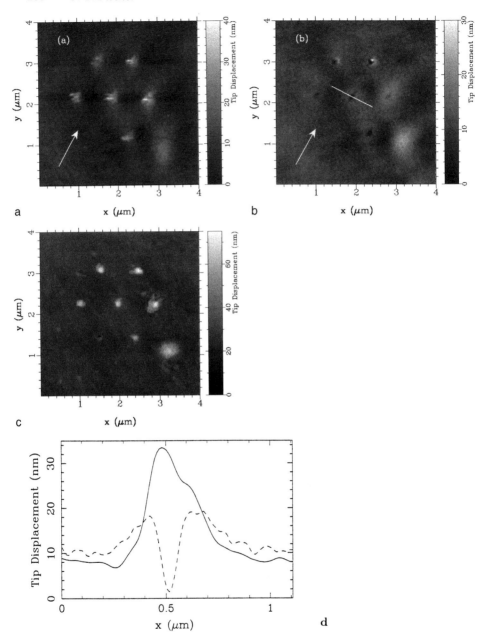

Fig. 7.14a–d. PSTM and AFM images of the lithographically produced rectangular glass post on a glass substrate. PSTM images are shown for two incident polarizations of the incident light: TM (**a**) and TE (**b**). The *arrows* indicate the direction of the projection of the incident wave vector parallel to the surface of the substrate. (**c**) AFM image; (**d**) Line profiles of PSTM images. The *cut line* in (**d**) follows the *white line* in (**b**). TM (*solid line*), TE (*dashed line*)

the fluorescence lifetimes as a function of the relative position of an aperture probe and the molecule.

An indicator of local evanescent fields of the SNOM probe, which are the basis of image formation of a near-field microscope, is often obtained from the dependence of the SNOM signal on distance from a dielectric reference surface. The decisive part of near-field interactions between probe and sample to contrast generation in near-field microscopy can be seen quite clearly in such dependencies for the case of resonantly excited SNOM probes. *Fischer* and *Pohl* [7.66] showed experimentally the tuning of the local plasmon resonance of a protrusion from a thin film of gold by varying the distance between the protrusion and a dielectric interface. Assuming a realistic model, *Dereux* et al. [7.128] were able to give an explanation for the experimentally observed effects by numerical calculations.

This near-field interaction which is related to near-field interactions in surface-enhanced spectroscopy [7.129] leads to mechanisms of contrast formation in near-field microscopy which are not accessible in other microscopic techniques. *Silva* and *Schultz* [7.130] made an attempt to exploit the enhanced sensitivity by using a resonantly excited silver nanoparticle as a probe for their Scanning Near-Field Optical Magnetic Kerr Microscope (SNOKE).

Taking into account near-field interactions between probe and sample complicates the interpretation of SNOM signals. In some cases simple quasistatic models are used for an interpretation of the data leading to an understanding of the main physical processes involved [7.48, 131, 132]. *Zenhausern* at al. [7.48] demonstrated the near-field interaction of a vibrating tip with a dielectric interface in their Scanning Interferometric Apertureless Microscope (SIAM), a configuration which is similar to the one employed by *Bachelot* et al. [7.32]. A schematic view of the SIAM configuration is shown in Fig. 7.15. The object and the tip are irradiated with a focused beam of light. The irradiating light induces coherently excited dipoles in both the tip and small features of the object. The scattered light interferes with the reflected light and an adjustable reference signal. A vibration of the tip induces a modulation of the interference signal. After subtraction of a "lift-off" signal, a remaining optical signal is obtained, which scales with the third power of distance between probing tip and the feature of the object. This optical signal, which is used to build up the image, is due to the interaction of the externally driven dipoles of the tip and the object feature. The "lift-off" signal is obtained as a signal at a reference site in the absence of the feature but the way in which the lift-off signal is obtained is not clearly disclosed. The near-field interaction of probe and sample is also assumed to play an important role in the configuration of *Koglin* et al. [7.132] who use the light emitting tetrahedral tip as a SNOM source. The tip is assumed to act as a radiating dipole excited from within the body of the tip. The probing dipole interacts with the surface of the sample inducing there a mirror dipole. The overlap of the dipole and the induced mirror dipole leads to a modified effective dipole of the tip which determines the radiative SNOM signal.

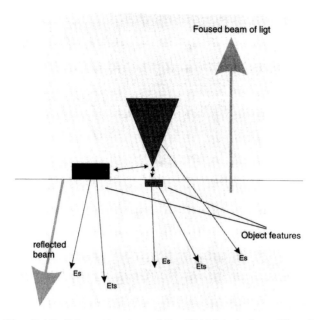

Fig. 7.15. Schematic view of the SIAM technique. The tip and the object is irradiated by a focused beam of light. The amplitudes of the electric field of (1) light reflected from the substrate (2) of scattered light E_s originating from the tip as well as from object features and (3) a contribution E_{ts} due to the interaction between the tip and the object features interfere to form the detector signal. E_{ts} varies strongly with the distance between tip and the feature, when the distance between the feature and the tip is small

7.6.2.5 Angular Spectroscopy. The angular distribution of light emitted from a small source is influenced by the presence of an interface and it strongly depends on the distance between source and sample. The angular radiation pattern of small light sources in front of an interface was investigated both experimentally and theoretically [7.125, 133, 134]. A prominent effect in this context is the excitation of evanescent modes, when the source dives into the range of evanescent modes of an interface between a medium of lower (air) and higher refractive index (glass). These modes are evanescent in the air and radiative in the glass. Within the glass, the modes propagate into the angular range of total internal reflection (Range III_2 of Fig. 7.4). A source approaching such a dielectric interface is invisible if observed from within the glass at angles within the angular range of total internal reflection, when the source is remote from the interface. It becomes visible, however, when the source penetrates the range of the evanescent modes and becomes increasingly brighter as the source approaches further within this range. The importance of the redistribution of the angular pattern for a illumination mode SNOM signal was pointed out in the context of near-field microscopy [7.12, 37, 135]. On the basis of this effect, the inverse PSTM configuration of a SNOM was suggested [7.37] and realized [7.63, 64]. *Dereux* and *Pohl* [7.135] emphasized that by analyzing the angular pattern of radiation of a probe in near-field interaction with a sample, interesting information can be obtained, e.g., about the lateral position of a sample detail with respect to the probe. *Hecht* et al. [7.136] realized an inverse PSTM configuration with the possibility to map the angular pattern and they investigated local surface plasmon excitation with such an instrument. The excitation of surface plasmons with a SNOM source leads to very selective angular radiation pattern. A further exploration of angular spectroscopy with a SNOM may well become a very interesting subject.

7.7 Spectroscopic Applications of SNOM

The most important property of SNOM in comparison to other forms of optical microscopy and to other forms of scanning probe microscopy is the potential of a high lateral resolution below the diffraction limit in conjunction with optical spectroscopic contrast. A few applications are described, where characteristic optical information can be obtained at a resolution in the 1–50 nm range, well below the diffraction limit. Most applications of SNOM were, however, up to now limited in resolution to the range of 200 to 50 nm. There are quite a few applications where new and important information can be obtained with such a resolution. In this respect, applications of SNOM to single molecule detection and spectroscopy, to the investigation of thin organic films and biological preparations, to semiconductors and optical waveguide structures are dealt with in this section.

7.7.1 SNOM at High Lateral Resolution

There are challenging optical problems to be addressed at a resolution in the 1–10 nm range. Among them are:

1) The investigation of locally resolved energy transport phenomena in photosynthesis and in aggregates of dye molecules.
2) The spatial resolution of the base sequence in single- or double-stranded DNA molecules.
3) The determination of spatial excitation profiles of extended quantum states and demonstration of the breakdown of selection rules for excitation with a near-field probe [7.137].
4) The correlation of material composition and magnetization at the surface of thin magnetic films.

With a few SNOM methods a resolution substantially below 10 nm was demonstrated. In all these cases no apertures were used as SNOM probes. SNOM images showing a resolution in the 3 nm range were in some cases obtained using the PSTM method [7.60, 61]. *Brückl* et al. [7.62] demonstrated

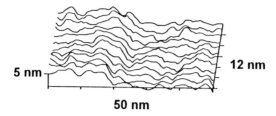

5 nm

50 nm

12 nm

Fig. 7.16. Detail of a PSTM image of the surface of a cleaved NaCl crystal showing a lateral resolution in the nm range. The full scan ranges in x, y and z direction are given [7.62]

in images of the surface of a cleaved NaCl crystal obtained with a PSTM configuration details of the surface relief structure of only 3 nm (Fig. 7.16). The potential for material contrast of this method was also demonstrated by vibrating the tip and using the modulated signal as an indicator of local variations of the refractive index of the surface. These researchers suggested to use this method for the imaging of submolecular details in macromolecules.

Specht et al. [7.87], *Zenhausern* et al. [7.48] and *Bachelot* et al. [7.32] also used external illumination schemes and opaque tips. *Zenhausern* et al. [7.48] obtained near-field images with their SIAM technique showing an edge resolution of 1 nm. The spectroscopic potential of the SIAM technique was demonstrated [7.138] by means of dye molecules bound to a tobacco mosaic virus (Fig. 7.17) or simply dye molecules dried onto piece of cleaved mica as an object. In this way, *Martin* et al. were able to detect only few molecules by a resonant scattering process and suggest that the detection of single molecules with this technique should be possible. The influence of the absorbing dye molecules for a wavelength close to the absorption maximum of the dye is an enhancement of scattering from the tip rather than an extinction. *Bachelot* et al. [7.138] applied a very similar technique in the mid infrared domain

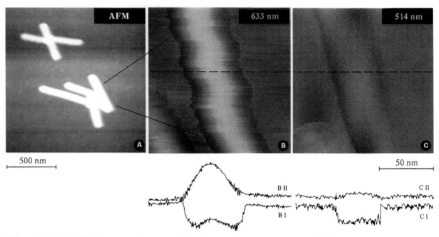

Fig. 7.17A–C. Image of stained Tobacco Mosaic virus. (**A**) AFM Image, (**B**) SIAM image at 633 nm, (**C**) at 514 nm. The line scans show signals before (I) and after (II) subtraction of the lift-off signals [7.138]

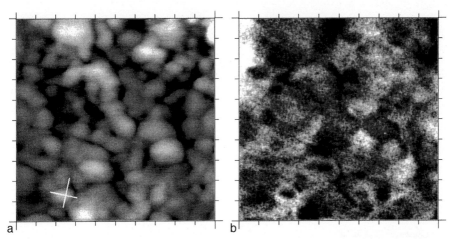

Fig. 7.18a, b. SNOM with the tetrahedral tip: (a) STM image of inverted mixed film of silver and gold, (b) simultaneous SNOM image. The scan range is 125 × 125 nm^2

(10.6 μm) and obtained images at a resolution in the order of 10 nm. Recently they were able to find an optical contrast for boron implanted regions in a substrate of silicon.

Using the tetrahedral probe, *Koglin* et al. [7.132] demonstrated material contrast in SNOM images with a lateral resolution in the 10–1 nm range also with an illumination mode of a SNOM. Figure 7.18 shows the SNOM image and the simultaneously recorded STM image of a mixed film of silver and gold. The film was prepared, as depicted schematically in Fig. 7.19. First a silver film of 0.4 nm nominal thickness was deposited onto a piece of freshly cleaved mica. A film of small unconnected silver islands is formed in this way. Then a 50 nm thick film of gold was deposited, such that the silver film is embedded into the surface of the gold film. The mixed film was then deposited upside down onto an ITO (Indium Tin Oxide) substrate. The surface of the gold film with the embedded silver grains was thus exposed and investigated in a transmission SNOM arrangement using the tetrahedral tip as a probe. In the STM image the topography of the mixed metal film is seen. Since the surface film consists of a replica of the atomically flat mica surface, the surface of this film is much smoother than the surface of a usual evaporated film. Grains are seen because the grains do not coalesce at the mica surface. In the SNOM image the grains appear as a distribution of dark and bright spots. In addition, some of the dark grains seem to be embedded in bright grains. Such a dark grain, as marked in the corresponding STM image is interpreted as a silver grain embedded in the surface of a gold grain. No topographic feature is seen in the simultaneously recorded STM image at the sites of a transition from the bright to the dark region. Line scans through the marked grain are shown in Fig. 7.20a,b for two orthogonal directions. The transition from bright to dark is limited to a width of less than 1 nm in this

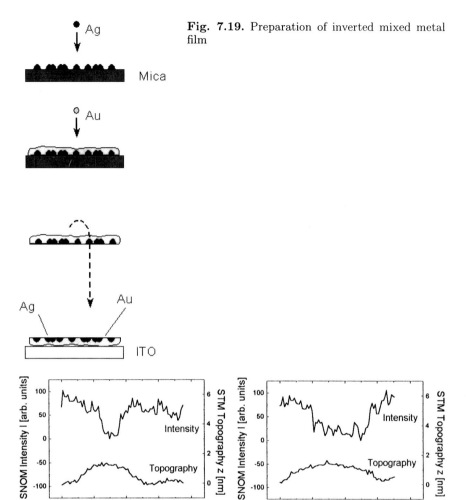

Fig. 7.19. Preparation of inverted mixed metal film

Fig. 7.20a, b. Line traces for cut lines indicated in Fig. 7.18a

case. It was shown that the difference in the complex dielectric constants of silver and gold at optical frequencies explains quite well the contrast between silver and gold grains observed in the images [7.132]. It is therefore expected that spectroscopic contrast of this method can be obtained reflecting the dependence on wavelength of the polarizability of the sample and of the tip.

The techniques described in this section have the potential to SNOM imaging at molecular lateral resolution and therefore are of interest in addressing the kind of problems mentioned in the beginning of this section.

7.7.2 Single-Molecule Detection

The field of Single-Molecule Spectroscopy (SMS) or of other nanoscopic particles by their luminescence has recently become an exciting new field of research [7.1]. New important information was obtained by studying the physico-chemical properties of individuals out of a large population of molecules. Single-molecule studies were first performed at low temperatures using conventional optics, but first clear microscopic images arising from the fluorescence of single molecules at room temperature were obtained by near-field microscopy by *Betzig* and *Chichester* [7.50]. In the mean time, it has become evident that spectroscopic imaging of single molecules is possible with conventional microscopy as well. The growing field of Single-Molecule Detection (SMD) and SMS has been extensively reviewed [7.24]. *Trautman* and *Ambrose* came to the following conclusions concerning near-field microscopy of single molecules [7.1]:

"Without delving further into the details, suffice it to say that to achieve a photon flux of 10^{20}–10^{21} cm^{-1}s^{-1} without heating the tip, it is necessary to use aperture diameters no smaller than $\lambda/5$ with the present generation of probes. The upshot of all this is that the kind of experiments performed on single molecules using the near field to date could have been done just as well with diffraction limited excitation. There remain however, the studies which are no doubt to come where the higher resolution of the near field will be necessary. It is in this realm, that near field should find its niche."

Apart from the first low temperature SMD studies by *Moerner* and *Kador* [7.140] who used the absorption of a single molecule as a signal, up to now all optical studies on single-molecule detection were performed using fluorescenc. There should also be a possibility to detect single molecules in near-field microscopy using the effect of a near to resonant excitation of a molecule on the scattering of the near-field probe as was pointed out by *Fischer* [7.44], who was able to detect an estimated number of 500 molecules in the near-field range of a resonantly excited aperture of 0.3 μm diameter in a silver film to change the scattering intensity of this hole by as much as 10 %. As mentioned above, *Martin* et al. [7.138] showed recently that with their near-field microscopic method they can detect a small number of light absorbing molecules and possibly single molecules by their effect on the scattering from the probing tip at a lateral resolution of 10 nm. A detection of single resonantly excited molecules by other means than their fluorescence should have some specific advantages for the investigation of single molecules. The absorption cross section should not be as sensitive to the vicinity of the probe as are the luminescent properties of a molecule. Furthermore, a new class of stable dye molecules which do not fluoresce could be investigated including photoreceptor molecules such as bacteriorhodopsin and the visual pigments [7.141], phytochromes [7.142] and the photosynthetic units [7.42].

In some SMD investigations with the near-field microscope, specific near-field effects become important. They are due to an interaction of the molecules with the probing tip. An example of such effects is the influence of the aperture probe on luminescence lifetime of a single molecule [7.126, 127]. An indication of other effects such as a tip induced Stark shift when applying a voltage across amolecule were observed by *Moerner* et al. [7.143] in a low temperature SNOM setup. Such influences could become a specific advantage of near-field microscopy, if it is possible to perturb single molecules by an interaction with a probing tip in a controlled way.

A large variety of spectroscopic techniques can be applied to the study of single molecules. By restricting the observed volume in fluorescence studies of dye molecules, it is possible to observe the emission properties of single molecules. Excitation and emission spectra of single molecules, which differ strongly from the averaged spectra, can be recorded. The time correlation of photons emitted from a single molecule allows to observe the dynamics of the fluorescence decay, spectral diffusion, rotational and translational diffusion and also of the statistical fluctuations due to chemical reactions or conformational changes of a molecule. The observation of single members out of a large population reduces the complexity of interpreting spectroscopic or kinetic data. Thus the fluorescence of a single molecule may be described by a single lifetime whereas the fluorescence decay of a larger population can only be represented by a distribution of lifetimes due to a variation of the local environment of single molecules. For obtaining useful spectroscopic information from single luminescent centers it is often important to perform experiments at low temperature. Only for these conditions variations in the spectroscopic properties between individual sites become discernible due to the strongly decreased linewidths.

One may expect that the niche of SMD studies by SNOM at very high lateral resolution can be filled by applying the technique of *Martin* et al. [7.138], the PSTM method or by using a new generation of probes such as, e.g., the tetrahedral probe. It should then be possible to localize and investigate the interaction of dye molecules with their environment on a lateral scale of 1 to 10 nm.

7.7.3 Thin Organic Films and Biological Preparations

There have been several examples of SNOM applied to thin organic films [7.144–147] and to biological preparations [7.22, 148–150] where details can be seen in SNOM images which cannot be seen in conventional light micrographs. In biological applications of a SNOM to the investigation of cell surfaces, the correlation of optical and topographic information can be a specific advantage of a SNOM. Figure 7.21 shows fluorescently labeled IGE receptor molecules to be distributed predominantly on hills of a cell surface, implying an optimal accessibility for external ligands [7.151]. This image showing a moderate resolution of about 150 nm was obtained using an uncoated tapered fiber probe in an internal reflection configuration. One may expect that SNOM should find further applications in the field of biology, especially

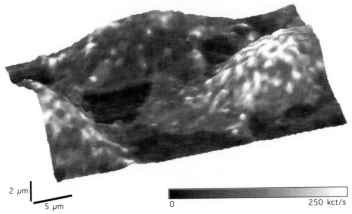

2 µm

5 µm

0   250 kct/s

Fig. 7.21. Rat basophilic leukemia cells (2H₃ line) with cy5-labeled IgE bound to its specific FCE surface receptor. Crosslinking by multivalent antigens of IgE on such cells is responsible for the immediate allergic response. The topography is represented by the 3D surface plot and the gray value shading reflects the intensities corresponding to the distribution of fluorescent IgE (*lighter shades* denote brighter signals). The receptor appears to be located predominantly on the "hills" and not in the "valleys", implying an optimal accessibility for the external ligands [7.151]

in combination with an operation in liquid environment [7.152]. Using an aperture-NSOM, *Higgins* and *Barbara* [7.153] made an effort to determine the range of exciton migration in *J*-aggregates of cyanine dyes. *J*-aggregates [7.154] are aggregates of dye molecules with a delocalised electronic excitation which have been extensively studied due to their importance in the sensitization of photographic films and as model systems for the light harvesting pigments of the photosynthetic apparatus. *Higgins* and *Barbara* found no indication of exciton migration beyond the resolution limit of 50 nm of their method and conclude that 50 nm should be an upper limit of exciton migration in their system. In the photosynthetic unit, exciton migration can be expected to occur only over a distance of maximally 30 nm, corresponding to the overall size of the photosynthetic unit including the light harvesting pigments [7.42]. A better resolution in SNOM has to be obtained to address the problems of locally resolved energy migration in such systems. It may be expected that a combination of SNOM at high resolution and SMD will open a wide field of applications to the investigation of artificially prepared systems which are of interest in the field of molecular biology. Recently, *Kramer* et al. [7.155] succeeded to apply SNOM to obtain images of the fluorescence of monomolecular films at the air water interface. For this purpose, the problem to approach the SNOM tip to only a few nm to the air–water interface is a difficult task that had to be solved. SNOM may now prove to be an important spectroscopic technique to study thin film formation, two dimensional crystallization processes and self assembly processes at the air water interface.

7.7.4 Semiconductors

The investigation of optoelectronic semiconductor devices has become an important field of applications for SNOM. Even at a moderate lateral resolution of about 100–300 nm, new and important information can be obtained by SNOM, which is not obtainable by other means. By using an external reflection mode aperture-NSOM suitable for cryogenic temperatures (down to 2 K) *Hess* et al. [7.156] were able to identify luminescent centers with sharp spectrally distinct emission lines in GaAs/AlGaAs quantum wells. They derive from effects of temperature and magnetic field on luminescence spectra and from linewidth measurements that the luminescent centers arise from excitons laterally localized at interface fluctuations. They state: "In sufficiently narrow wells, nearly all emission originates from such centers. Near-field microscopy and spectroscopy provide a means to access energies and homogeneous linewidths for the individual eigenstates of these centers, and thus opens a rich area of physics involving quantum resolved systems." Such quantum dot luminescent centers have also been identified by conventional microscopy [7.157]. However, the resolution with conventional microscopy in the wavelength range above 630 nm without immersion optics is limited to about 1 µm compared with a resolution in the order of 100 to 300 nm as obtained with the aperture-NSOM. Meanwhile more detailed investigation of single quantum dot spectra was performed by *Gammon* et al. [7.158] using a non scanning near-field method similar to the one described previously by *Fischer* [7.44] using a fixed aluminum mask with small holes of defined diameters between 0.2 and 25 µm in contact with a GaAs/AlGaAs quantum well as near-field probes.

Further near-field photoluminescence studies were performed to investigate the photoluminescence of porous silicon [7.159] and to investigate the uniformity of semiconductor surfaces [7.160, 161].

Quite a few papers deal with the investigation of microfabricated optoelectronic semiconductor devices by aperture-NSOM: With external and – partly – also with internal reflection mode SNOM instruments the following problems have been addressed:

1) Near-field optical spectroscopy of quantum wires [7.162, 163].
2) "Spatially and spectrally resolved imaging of quantum-dot structures using near-field optical technique" [7.164].
3) Spatially resolved photoluminescence spectroscopy of lateral *pn*-junctions prepared by Si-doped GaAs [7.165].

Furthermore, light emitting semiconductor devices were investigated with the aperture-NSOM by simply using the probe as a detector [7.166, 167].

Finally, near-field photocurrent spectroscopy is another tool for the investigation of optoelectronic devices [7.73, 168–172].

In order to give an example of the wealth of information, which can be obtained by near-field photoluminescence studies from optoelectronic semiconductor devices, we report in some more detail about a recent investigation by *Richter* et al. [7.172] on the photoluminescence of a single GaAs quantum ridge, a special form of a quantum wire (QWR) as fabricated by *Noetzel* et al. [7.173].

Photoluminescence investigations can be performed in different ways. In the external reflection mode, for obtaining photoluminescence emission spectra (PL), the sample is excited locally with the SNOM probe and the spectrally resolved global photoemission is measured as shown schematically in Fig. 7.22. Alternatively, for obtaining photoluminescence excitation spectra (PLE), the excitation wavelength of the exciting laser is scanned and the photoluminescence is measured at a fixed wavelength. Luminescence spectra may also be obtained in a collection mode, where the luminescence is excited globally, whereas the luminescence is detected locally by the probe. *Richter* et al. used several techniques and obtained important complementary information. Figure 7.23 shows a set of PL emission spectra (x-axis) with spatially resolved excitation along a cross section perpendicular to the QWR for an excitation energy of 1.959 eV. Figure 7.24 exhibits a set of PLE excitation spectra with spatially resolved excitation for an emission at 1.46 eV, and Fig. 7.25 shows a set of PL emission spectra with spatially resolved excitation at 1.5 eV. From an investigation of such spatially resolved spectra, detailed information about the electronic structure of the quantum wire and its interaction with the QW have been obtained. From the rather large spatial range of luminescence for energies at 1.465 eV – characteristic for the emission from the QWR – in the PL spectra of Fig. 7.23 *Richter* et al. deduced a diffusion

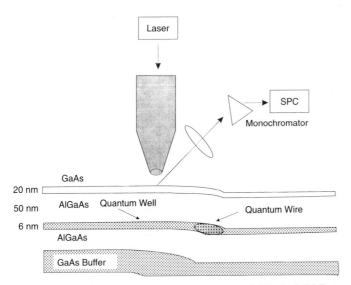

Fig. 7.22. Scheme of SNOM setup to record PL and PLE spectroscopic SNOM images and schematic view of quantum wire, which is embedded into a quantum well

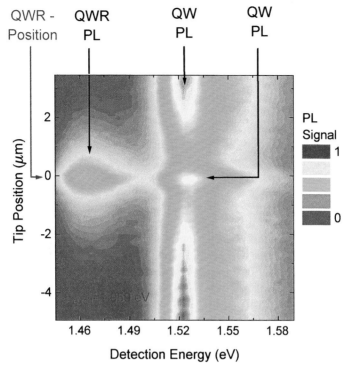

Fig. 7.23. Spatially resolved near-field photoluminescence spectrum of the quantum wire heterostructure. The spectrum was recorded for spatially resolved excitation of the sample at 1.959 eV. The tip was scanned along the lateral direction perpendicular to the wire. The PL intensity (in arbitrary units) is plotted as a function of tip position and detection energy. The color red corresponds to high, while purple corresponds to low PL intensity. The quantum wire emission is centered around 1.46 eV. Note that, in addition to the flat quantum well luminescence at 1.522 eV, a further, slightly blue shifted, sidewall quantum well emission is resolved

of carriers excited in the QW into the QWR where they recombine. The PLE spectra of Fig. 7.24 with a fixed emission energy of 1.46 eV, which is characteristic for the emission of the QWR, show that electrons excited in the quantum well at excitation energies above 1.5 eV diffuse into the quantum wire. The spectrum shows a depletion of the luminescence in regions adjacent to the QWR probably due to a reduced absorption of the deformed QW in these regions. In the PL spectrum of Fig. 7.25 recorded with a fixed excitation energy of 1.5 eV which can excite the QWR but not the QW, one can notice that excitation of the QWR leads to photoemission at a higher energy which is explained by carrier diffusion from the quantum wire to the quantum well with a concomitant thermal activation leading to a higher emission energy.

In summary, the application of SNOM to optoelectronic semiconductor devices has become an important application of near-field microscopy. New

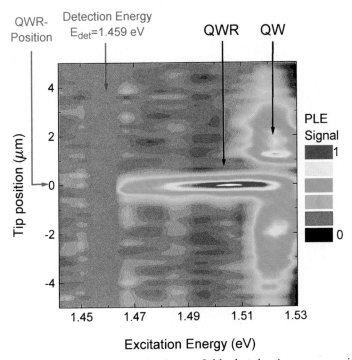

Fig. 7.24. Spatially resolved near-field photoluminescence excitation spectrum. The PLE signal is plotted as a function of tip position and excitation energy. The scan direction is perpendicular to the wire. The photoluminescence was detected at 1.46 eV. The color red corresponds to high and purple to low intensity

information can be gained even at a moderate resolution in the order of 100–300 nm, which is not accessible by other methods.

7.7.5 Waveguide Structures

By using the PSTM method, $LiTaO_3$ waveguides were evaluated by measuring the intensity profile of the evanescent field of propagating modes [7.113]. The measured profile yields information about scattering sources and about the refractive index profile. The PSTM method is expected to be widely used in the high spatial resolution diagnostics of waveguide structures in the future.

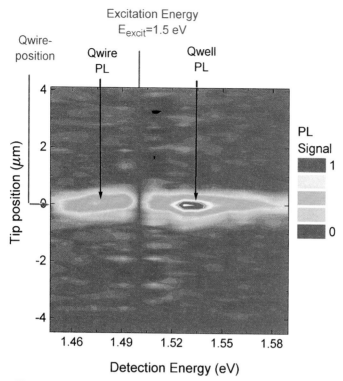

Fig. 7.25. Near-field PL spectrum for resonant, spatially resolved quantum wire excitation at 1.5 eV. The PL signal is plotted as a function of detection energy and of tip position along a profile perpendicular to the wire

7.8 Nanophotolithography

As most other microscopes, a near-field optical microscope not only can be used to obtain images but also as a tool to modify a sample at a local scale. Nanophotolithography can be performed with illumination mode SNOM configurations. Nanophotolithography of photoresists with a SNOM was demonstrated by *Betzig* and *Trautman* [7.22] and subsequently by *Krausch* et al. [7.174] and by *Naber* et al. [7.175]. Figure 7.26 shows an AFM image of a developed photoresist structure as written with an aperture-NSOM [7.175]. Figure 7.26a displays a line profile across a written structure. The resolution is about 80 nm and the depth of the developed structure is only about 20 nm. The resolution of the written structure deteriorates with longer exposures. Therefore, one cannot expect that the nanolithographic process with a SNOM is compatible with conventional photolithography, which needs an irradiation through the full thickness of the resist of 50 nm in order to be able to use the resist as an etch resist for further processing. Nanophotolithography is therefore not directly compatible with conventional photolithography. Much

(a)

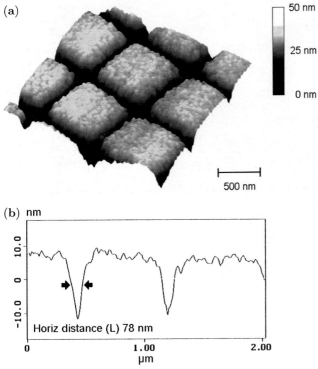

50 nm

25 nm

0 nm

500 nm

(b) nm

Horiz distance (L) 78 nm

0   1.00   2.00
                     µm

Fig. 7.26. (a) AFM image of a developed photoresist, on which a grid pattern was exposed by an aperture-NSOM. (b) Section analysis of the two lines. The aspect ratio of width to depth is about 4 : 1

thinner photosensitive layers have to be used such as monomolecular layers and it is doubtful whether they can be used as etch resists. By modifying a very thin layer or a surface, it is possible, however, to alter the chemical reactivity of a surface or its adsorption properties. Such lithographic processes on the basis of monomolecular films are of potential interest to create surface structures of differing chemical reactivity on a nanoscopic scale as demonstrated experimentally in connection with electron beam lithography by *Zingsheim* [7.176] and in connection with scanning probe microscopy by *Calvert* [7.177]. A chemical surface structure is then used as a seed to build up nanostructures by specific adsorption to the lithographically defined locations. First attempts to modify monolayers of a dye by a near-field microscope were performed by *Jiang* et al. [7.178].

Scanning probe microscopy, in general, is a useful tool to create nanostructures, as shown by *Mamin* and *Rugar* [7.179, 180], for example. They used a heated AFM tip to write thermally induced nanostructures into a polymer film, where the heating of the tip could be performed either by irradiation with a focused laser beam or by resistive heating. *Jersch* and *Dickmann* [7.181] performed optically induced surface modification using the effect of

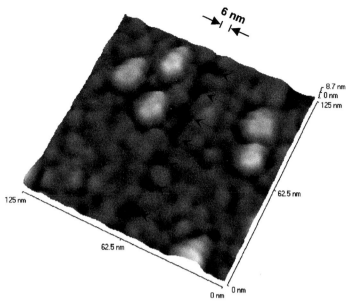

Fig. 7.27. AFM image of pit structures created by the FOLANT technique (non contact mode) on a gold surface [7.182]

an opaque tip illuminated by laser pulses. The researchers interpreted the effects of light pulses on the samples as being related to a nonlinear effect of a highly localized optical near field of the irradiated tip. An example of a remarkably well defined nanostructure written into a thin metal film by the FOLANT (FOcusing of LAser radiation in Near-field of a Tip) technique in combination with a non contact mode of operation of an AFM [7.182] is depicted in Fig. 7.27.

In the context of nanofabrication by scanning probe techniques, a replication of such structures by other techniques is of interest. *Terris* et al. [7.183] introduced a method of replicating surface relief structures which were written with an atomic force microscope.

Contact imaging by energy transfer [7.8], as mentioned above, is a rather unique form of replicating nanostructures. The method of contact imaging has not been yet developed further. In combination with the tools of scanning probe microscopy now available, it could become a powerful method for pattern transfer using nanostructures as a mask which could be produced by the methods of scanning probe microscopy, electron beam lithography or by other means such as "natural lithography" using latex spheres or aggregates of latex spheres as a mask [7.184, 185]. Again, an essential aspect in this scheme is the use of very thin monomolecular films as medium to fabricate surface structures of differing chemical reactivity as mentioned above. On a larger scale such a photochemical patterning of a surface is already of interest in a new technique for gene sequencing introduced by *Pease* et al. [7.186]. *Palacin* et al. [7.187] introduced the related method of microstamping.

7.9 Conclusions

Scanning near-field optical microscopy is a form of scanning probe microscopy, which makes use of the optical interaction between a nanoscopic tip and a sample to obtain spatially resolved optical information, which is not bound to the diffraction limit. The development of aperture-NSOM using tapered metal coated monomode fibers with an aperture at the apex as a probe and shear force microscopy as a technique to control the distance between tip and sample during scanning triggered many efforts to apply the SNOM technique to a broad variety of samples concerning the investigation of semiconductor surfaces, optoelectronic devices, waveguide structures, thin organic films and biological samples. In most applications the resolution of SNOM is limited to 200–50 nm. The simultaneous optical and topographic information is an important advantage of the SNOM method even at moderate resolution compared to high-resolution conventional light microscopy. SNOM at moderate resolution is also of interest for experimental conditions where the fundamental diffraction limit cannot be reached by classical light microscopy due to technical reasons, e.g., at low temperature where immersion optics cannot be used. Emerging microfabricated probes may trigger a more widespread application of the technique. Near-field microscopy in the mid infrared spectral domain should become very relevant due to the specific chemical information which can be obtained and because no other microscopic methods are available in this spectral range. Recent developments of SNOM probes without apertures open a perspective of SNOM at a resolution in the 1–10 nm range. It is expected that first applications of these methods will soon emerge and thus open new perspectives for the SNOM technique. Due to the specific optical interaction with a sample a unique potential of SNOM can be seen in:

1) The spectroscopic characterisation at a molecular scale.
2) The control and modulation of the spectroscopic properties of nanostructures and of single molecules by the interaction with a probing tip.
3) The controlled modification at a nanoscopic scale due to specific photochemical and other optical interactions.

Apart from the aspect of microscopy, the study of optical phenomena at nanoscopic dimensions can profit from SNOM. The design of SNOM probes is intimately related to the understanding of light propagation, confinement and compression by electromagnetic fields at interfaces.

Acknowledgements

I would like to acknowledge the support I obtained in writing this review by fruitful discussions with my colleagues, H. Danzebrink J. Ferber, H. Fuchs, W. Göhde, J. Koglin, A. Naber. I would like to thank for providing generously background material and illustrations to: P.F. Barbara, H. Brückl, A. Dereux, J. Jersch, T. Jovin, K. Karrai, A. Kirsch, R. Kopelman, K. Lieberman, C. Lienau, N. Naya, W. Noell., C. Obermüller, E. Oesterschulze, M. Ohtsu, D. Pohl, X. Sunney Xie, J.C. Weeber, K. Wickramasinghe. I am indebted to R. Reichelt for the SEM image of the tetrahedral tip. Last but not least I thank J. Heimel and B. Gotsman for critical reading of the manuscript.

References

[7.1]   T. Basché, W.E. Moerner, M. Orrit, U.P. Wild (eds.): *Single-Molecule Optical Detection, Imaging and Spectroscopy* (VCH, Wiley, New York 1997)
[7.2]   T. Funatsu, Y. Harada, M. Tokunaga, K. Saito, T. Yanagida: Nature **374**, 558 (1994)
[7.3]   W. Denk, J.H. Strickler, W.W. Webb: Science **248**, 73 (1990)
[7.4]   P.E. Hänninen, S.W. Hell, J. Salo, E. Soini: Appl. Phys. Lett. *66*, 1698 (1995)
[7.5]   J.D. Jackson: *Classical Electrodynamics*, 2nd edn. (Wiley, New York 1975)
[7.6]   E.H. Synge: Philos. Mag. **6**, 356 (1928)
[7.7]   H. Kuhn: On possible ways of assembling simple organised systems of molecules, in: *Structural Chemistry and Molecular Biology*, ed. by A. Rich, N. Davidson (Freeman, San Francisco 1968) pp. 566–571
[7.8]   U.C. Fischer, H.P. Zingsheim: Appl. Phys. Lett. **40**, 195 (1982)
[7.9]   E.A. Ash, G. Nichols: Nature **237**, 510 (1972)
[7.10] D.W. Pohl, W. Denk, M. Lanz: Appl. Phys. Lett. **44**, 651 (1984)
[7.11] Talk given by H. Kuhn at the second Meeting of Molecular Electronic Devices (1983) with a report about results of U. Fischer on scanning near-field optical microscopy. See H. Kuhn: In *Molecular Electronic Devices II*, ed. by F.L. Carter (Dekker, New York 1987) p. 415
[7.12] U.C. Fischer: J. Vac. Sci. Technol. B **3**, 386 (1985)
[7.13] A. Lewis, M. Isaacson, A. Harootunian, A. Muray: Ultramicrosc. **13**, 227 (1984)
[7.14] U. Dürig, D.W. Pohl, F. Rohner: J. Appl. Phys. **59** (10), 3318 (1986)
[7.15] A. Harootunian, E. Betzig, M. Isaacson, A. Lewis: Appl. Phys. Lett. **49**, 674 (1986)
[7.16] E. Betzig, J.K. Trautman, T.D. Harris, J.S. Weiner, R.L. Kostelak: Science **251**, 1468 (1991)
[7.17] E. Betzig, P.L. Finn, J.S. Weiner: Appl. Phys. Lett. **60**, 2484 (1992)
[7.18] P.C. Yang, Y. Chen, M. Vaez-Iravani: J. Appl. Phys. **71**, 2499 (1992)
[7.19] D.W. Pohl: Nano-optics and scanning near-field optical microscopy, in: *Scanning Tunneling Microscopy II*, ed. by R. Wiesendanger, H.J. Güntherodt, 2nd edn., Springer Ser. Surf. Sci., Vol. 28 (Springer, Berlin, Heidelberg 1995) pp. 233-271
[7.20] D. Courjon, C. Bainier: Rep. Prog. Phys. **57**, 989 (1994)
[7.21] M. Paesler, P. Moyer: *Near-Field Microscopy* (Wiley, New York 1996)

[7.22]   E. Betzig, J.K. Trautman: Science **257**, 189 (1992)
[7.23]   M. Ohtsu: J. Lightwave Technol. **13**, 1200 (1995)
[7.24]   X. Sunney Xie, R.X. Bian, R.C. Dunn: Near-Field Microscopy and Spectroscopy of Single Molecules, Single Proteins and Biological Membranes, in *Focus on Multidimensional Microscopy*, Vol. 1, ed. by P.C. Chen, P.P. Hwang, H. Kim (World Scientific, Singapore 1997)
[7.25]   U.C. Fischer, J. Koglin, A. Naber, A. Raschewski, R. Tiemann, H. Fuchs: Near-Field Optics and Near-Field Optical Microscopy, in *Quantum Optics of Confined Systems*, ed. by M. Ducloy, D. Bloch (Kluwer, Dordrecht 1996) p. 309
[7.26]   D. Van Labeke, D. Barchiesi: Theoretical problems in Scanning Near-Field Optical Microscopy, in *Near Field Optics*, ed. by D.W. Pohl, D. Courjon (Kluwer, Dordrecht 1993) pp. 157–178
[7.27]   C. Girard, A. Dereux: Rep. Progr. Phys. **59** 657 (1996)
[7.28]   D.W. Pohl, D. Courjon (eds.): *Near Field Optics*, NATO ASI Ser. E. Appl. Sci., Vol. 242 (Kluwer, Dordrecht 1993)
[7.29]   M. Isaacson, M. Paesler (guest eds.): *Near Field Optics II*, Ultramicrosc. **57** (1995)
[7.30]   M. Paesler, N. van Hulst (guest eds.): *Near Field Optics III*, Ultramicrosc. **61** (1–4) (1995)
[7.31]   F. Zenhausern, M.P. O'Boyle, H.K. Wickramasinghe: Appl. Phys. Lett. **65**, 1623 (1994)
[7.32]   R. Bachelot, P. Gleyzes, A.C. Boccara: Optics Lett. **20**, 1924 (1995)
[7.33]   L.Novotny, D.W. Pohl, B. Hecht: Ultramicroscopy **61**, 1 (1995)
[7.34]   F. Keilmann: German Patent DE 3837389 (1988)
[7.35]   M. Fee, S. Chu, T.W. Hänsch: Optics Commun. **69**, 219 (1989)
[7.36]   U.C. Fischer: German Patent application DE 3916047 (1989)
[7.37]   U.C. Fischer, M. Zapletal: Ultramicrosc. **42-44**, 393 (1991)
[7.38]   C.W. McCutchen, Scanning **17**, 15 (1995)
[7.39]   G.F. Taylor: Phys. Rev. **23**, 655 (1924)
[7.40]   K. Lieberman, S. Harush, A. Lewis, R. Kopelman: Science **24**, 59 (1990)
[7.41]   W. Tan, D. Birnbaum, C. Harris, R. Merlin, B. Orr, Zhong-You Shi, S. Smith, B.A. Thorsrud, R. Kopelman: Near-field optics: chemical sensors, photon supertips and subwavelength spectroscopy, in *Michrochemistry, Spectroscopy and Chemistry in Small Domains*, ed. by H. Masuhara et al. (Elsevier, Amsterdam 1994) pp. 301–318
[7.42]   R.J. Codgell, P.K. Fyfe, S.J. Barrett, S.M. Prince, A.A. Freer, N.W. Isaacs, P. McGlynn, C.N. Hunter: Photosynthesis Res. **48**, 55 (1996)
[7.43]   J. Wessel: J. Opt. Soc. Am. B **2**, 1538 (1985)
[7.44]   U. Ch. Fischer: J. Opt. Soc. Am. B **3**, 1239 (1986)
[7.45]   U.C. Fischer, D.W. Pohl: Phys. Rev. Lett. **62**, 458 (1989)
[7.46]   D. Courjon, C. Bainier, F. Baida, C. Girard: Ultramicrosc. **61** (1–4), 117 (1995)
[7.47]   E. Betzig, S.G. Grubb, R.J. Chichester, D.J. DiGiovanni, J.S. Weiner: Appl. Phys. Lett. **63**, 3550 (1993)
[7.48]   F. Zenhausern, Y. Martin, H.K. Wickramasinghe: Science **269**, 1083 (1995)
[7.49]   H.A. Bethe: Phys. Rev. **66**, 163 (1944)
[7.50]   E. Betzig, J. Chichester: Science **262**, 1422 (1993)
[7.51]   C. Obermüller, K. Karrai: Appl. Phys. Lett. **67**, 3408 (1995)
[7.52]   J. Koglin, U.C. Fischer, H. Fuchs: J. Biomed. Opt. **1**, 75 (1996)
[7.53]   K. Karrai, R.D. Grober: Appl.Phys. Lett. **66**, 1842 (1995)
[7.54]   M.J. Gregor, P.G. Blome, J. Schofer, R.G. Ulbrich: Appl. Phys. Lett. **68**, 307 (1996)
[7.55]   C. Durkan, I.V. Shvets: J. Appl. Phys. **80**, 5659 (1996)

[7.56] W. Göhde, J. Tittel, Th. Basché, C. Bräuchle, U.C. Fischer, H. Fuchs: Rev. Sci. Instr. **68**, 2466 (1997)

[7.57] A. Jalocha, M.H.P. Moers, A.G.T. Ruiter, N.F. van Hulst: Ultramicrosc. **61**, 221 (1995)

[7.58] K. Lieberman, A. Lewis, G. Fish, S. Shalom, T.M. Jovin, A. Schaper, S.R. Cohen: Appl. Phys. Lett. **65**, 648 (1994)

[7.59] T. Ataka, H. Muramatsu, K. Nakajima, N. Chiba, K. Homma, M. Fujihira: Thin Solid Films **73**, 154 (1996)

[7.60] D. Courjon, K. Sarayeddine, M. Spajer: Opt. Commun. **71**, 23 (1989)

[7.61] R.C. Reddick, R.J. Warmack, T.L. Ferrell: Phys. Rev. B **39**, 767 (1989)

[7.62] H. Brückl, H. Pagnia, N. Sotnik: Scanning **17**, 24 (1995)

[7.63] B. Hecht, H. Heinzelmann, D.W. Pohl: Ultramicrosc. **57**, 228 (1995)

[7.64] U.C. Fischer, J. Koglin, H. Fuchs: J. Microsc. **176**, 281 (1994)

[7.65] U.C. Fischer, U.T. Dürig, D.W. Pohl: Appl. Phys. Lett. **52**, 249 (1988)

[7.66] U.C. Fischer, D.W. Pohl: Phys. Rev. Lett. **62**, 458 (1989)

[7.67] M. Spajer, D. Courjon, K. Sarayeddine, A. Jalocha, J.M. Vigoureux: Journ. Phys. III **1**, 1 (1991)

[7.68] J.A. Cline, H. Barshatzky, M. Isaacson: Ultramicrosc. **38**, 299 (1991)

[7.69] T. Kataoka, K. Endo, Y. Oshikane, H. Inoue, K. Inagaki, Y. Mori, H. An, O. Kobayakawa, A. Izumi: Ultramicrosc. **63**, 219 (1996)

[7.70] B. Hecht, H. Bielefeldt, Y. Inouye, D.W. Pohl, L. Novotny: J. Appl. Phys. **81**, 2492 (1997)

[7.71] G.A. Valaskovic, M. Holton, G.H. Morrison: Appl. Opt. **34**, 1215 (1995)

[7.72] C. Obermueller, K. Karrai, G. Kolb, G. Abstreiter: Ultramicrosc. **61** (1–4), 171 (1995)

[7.73] K. Karrai, G. Kolb, G. Abstreiter, A. Schmeller: Ultramicrosc. **61** (1–4) 299 (1995)

[7.74] T. Pangaribuan, K. Yamada, S. Jiang, H. Ohsawa, M. Ohtsu: Jap. J. Appl. Phys., Part 2 (Letters) **31**, AL1302 (1992)

[7.75] T. Pangaribuan, S. Jiang, M. Ohtsu: Electron. Lett. **29**, (1993)

[7.76] T. Pangaribuan, S. Jiang, M. Ohtsu: Scanning **16**, 362 (1994)

[7.77] T. Saiki, S. Mononobe, M. Ohtsu: Appl. Phys. Lett. **68** (19), 2612 (1996)

[7.78] D. Zeisel, S. Nettesheim, B. Dutoit, R. Zenobi: Appl. Phys. Lett. **68** (18), 2491 (1996)

[7.79] H. Raether: *Surface Plasmons on Smooth and Rough Surfaces and on Gratings*, Springer Tracts Mod. Phys., Vol. 111 (Springer, Berlin 1988)

[7.80] A. Sommerfeld: *Vorlesungen über theoretische Physik III. Elektrodynamik* (Dietrichsche Verlagsbuchhandlung, Wiesbaden 1948)

[7.81] L. Dobrzynski, A.A. Maradudin: Phys. Rev. B **6**, 3810 (1972)

[7.82] Ph. Lambin, T. Laloyaux, P.A. Thiry, J.P. Vigneron, A.A. Lucas: Europhys. Lett. **2**, 409 (1986)

[7.83] J. Ferber, U.C. Fischer, J. Koglin, H. Fuchs: Proc. SPIE Int. Soc. Opt. Eng. (USA), Vol. 2782, 535 (1996)

[7.84] D. Courjon, J. M. Vigoureux, M. Spajer, K. Sarayeddine, S. Leblanc: Appl. Opt. **29**, 3734 (1990)

[7.85] A. Jalocha, N.F. van Hulst: J. Opt. Soc. Am. B **12** (9), 1577 (1995)

[7.86] U.Ch. Fischer: Resolution and contrast generation in scanning near field optical microscopy, in *Scanning Tunneling Microscopy and Related Methods*, ed by R.J. Behm, N. Garcia, H. Rohrer (eds.) (Kluwer, Dordrecht 1990) pp. 475–496

[7.87] M. Specht, J.D. Pedarnig, W.M. Heckl, T.W. Hänsch: Phys. Rev. Lett. **68**, 476 (1992)

[7.88] C.B. Prater, P.K. Hansma, M. Tortonese, C.F. Quate: Rev. Sci. Instrum. **62**, 2634 (1991)

[7.89] S. Muenster, S. Werner, C. Mihalcea, W. Scholz, E. Oesterschulze: J. Microscopy **186**, 17 (1997); see also: S. Werner, O. Rudow, L. Mihalcea, E. Oesterschulze: Appl. Phys. A (in press)

[7.90] A.G.T. Ruiter, M.H.P. Moers, N.F. van Hulst, M. Deboer: J. Vac. Sci. Technol. B **14**, 597 (1996)

[7.91] N.F. van Hulst, M.H.P. Moers, B. Bölger: J. Microsc. **171**, 95 (1993)

[7.92] W. Noell, M. Abraham, K. Mayr, A. Ruf, J. Barenz, O. Hollricher, O. Marti, P. Güthner: Appl. Phys. Lett. **70** 1236 (1997)

[7.93] C.C. Williams, H.K. Wickramasinghe: Appl. Phys. Lett. **49**, 1587 (1986)

[7.94] K. Lieberman, S. Harush, A. Lewis, R. Kopelman: Science **24**, 59 (1990)

[7.95] H. Goettlich, W.M. Heckl: Ultramicrosc. **61**, 145 (1995)

[7.96] S. K. Sekatskii, V.S. Letokhov: Appl. Phys. B **63**, 525 (1996)

[7.97] A. Lewis, K. Lieberman: Nature **354**, 214 (1991)

[7.98] W. Tan, Z. Shi, S. Smith, D. Birnbaum, R. Kopelman: Science **258**, 778 (1992)

[7.99] N. Kuck, K. Lieberman, A. Lewis, A. Vecht: Appl. Phys. Lett. **61**, 139 (1992)

[7.100] H.U. Danzebrink, U.C. Fischer: in *Near Field Optics*, ed by D.W. Pohl, D. Courjon (Kluwer, Dordrecht 1993) pp. 303–308

[7.101] H.U. Danzebrink: J. Microsc. **176**, 276 (1994)

[7.102] H.U. Danzebrink, G. Wilkening, O. Ohlsson: Appl. Phys. Lett. **67**, 1981 (1995)

[7.103] H.U. Danzebrink: "Nahfeldmikroskopischer Sensor mit interner Signalwandlung"; Dissertation, Technische Universität Braunschweig (1997); PTB-Bericht F (27)

[7.104] R.C. Davis, C.C. Williams, P. Neuzil: Appl. Phys. Lett. **66**, 2309 (1995)

[7.105] R.C. Davis, C.C. Williams: Appl. Phys. Lett. **69**, 1179 (1996)

[7.106] S. Akamine, H. Kuwano, H. Yamada: Appl. Phys. Lett. **68**, 579 (1996)

[7.107] J.K. Trautman, E. Betzig, J.S. Weiner, D.J. DiGiovanni, T.D. Harris, F. Hellman, E.M. Gyorgy: J. Appl. Phys. **71**, 4659 (1992)

[7.108] M.H.P Moers, N.F. van Hulst, A.G.T. Ruiter, R. Bolger: Ultramicrosc. **57**, 298 (1995)

[7.109] D.A. Smith, S. Webster, M. Ayad, S.D. Evans, D. Fogherty, D. Batchelder: Ultramicrosc. **61**, 247 (1995)

[7.110] C.L. Jahncke, H.D. Hallen, A.M. Paesler: J. Raman Spectrosc. **27**, 579 (1996)

[7.111] D. Zeisel, B. Dutoit, V. Deckert, T. Roth, R. Zenobi: Anal. Chem. **69**, 749 (1997)

[7.112] R. Carminati, J.J. Greffet: Opt. Lett. **21**, 1295 (1996)

[7.113] Y. Toda, M. Ohtsu: IEEE Photonics Techn. Lett. **7**, 84 (1995)

[7.114] O. Marti, H. Bielefeldt, B. Hecht, S. Herminghaus, P. Leiderer, J. Mlynek: Opt. Commun. **96**, 225 (1993)

[7.115] J.C. Weeber, E. Bourillot, A. Dereux, J.P. Goudonnet, Y. Chen, C. Girard: Phys. Rev. Lett. **77**, 5332 (1996)

[7.116] T. Saiki, M. Ohtsu, K. Jang, W. Jhe: Opt. Lett. **21**, 674 (1996)

[7.117] N. Naya, S. Mononobe, R.U. Maheswari, T. Saiki, M. Ohtsu: Opt. Commun. **124**, 9. (1996)

[7.118] P. Dawson, F. de Fornel, J. P. Goudonnet: Phys. Rev. Lett. **72**, 2927 (1994)

[7.119] R.B.G. de Hollander, N.F. van Hulst, R.P.E. Kooyman: Ultramicrosc. **57**, 263 (1994)

[7.120] D.P. Tsai, Y. Wang, M. Moskovits, V.M. Shalaev, J.S. Sheh, R. Botet: Phys. Rev. Lett. **72**, 4149 (1994)

[7.121] B. Hecht, H. Bielefeldt, L. Novotny, Y. Inouye, D.W. Pohl: Phys. Rev. Lett. **77**, 1889 (1996)

[7.122] S.I. Bozhevolnyi, I.I. Smolyaninov, A.V. Zayats: Phys. Rev. B **51**, 17916 (1995)
[7.123] J.R. Krenn, W. Gotschy, D. Somitsch, A. Leitner, F.R. Aussenegg: Appl. Phys. A **61**, 541 (1995)
[7.124] D. Zeisel, B. Dutoit, V. Deckert, T. Roth, R. Zenobi: Anal. Chem. **69**, 749 (1997)
[7.125] K.H. Drexhage: *Progress in Optics* (North Holland, Amsterdam 1974) p. 165
[7.126] J.K. Trautman, J.J. Macklin: Chem. Phys. **205**, 221 (1996)
[7.127] R.X. Bian, R.C. Dunn, X. Sunney Xie: Phys. Rev. Lett. **75**, 4772 (1995)
[7.128] A. Dereux, J.P. Vigneron, P. Lambin, A.A. Lucas: Physica B **175**, 65 (1991)
[7.129] H. Metiu: Progr. Surf. Sci. **17**, 153 (1984)
[7.130] T.J. Silva, S. Schultz, D. Weller: Appl. Phys. Lett. **65**, 658 (1994)
[7.131] H. Hori: *Near Field Optics*, ed. by D.W. Pohl, D. Courjon (Kluwer, Dordrecht 1993) pp. 105–114
[7.132] J. Koglin, U.C. Fischer, H. Fuchs: Phys. Rev. B **55**, 7977 (1997)
[7.133] P. Fröhlich: Ann. Phys. (Leipzig) **65**, 577 (1921)
[7.134] C.K. Carniglia, L. Mandel, K.H. Drexhage: J. Opt. Soc. Am. **62**, 479 (1972)
[7.135] A. Dereux, D.W. Pohl: in *Near Field Optics*, ed. by D.W. Pohl, D. Courjon (Kluwer, Dordrecht 1993) p. 189
[7.136] B. Hecht, H. Bielefeldt, L. Novotny, Y. Inouye, D.W. Pohl: Phys. Rev. Lett. **77**, 1889 (1996)
[7.137] K. Cho, Y. Ohfuti, K. Arima: Surf. Sci. **363**, 378 (1996)
[7.138] Y. Martin, F. Zenhausern, H.K. Wickramasinghe: Appl. Phys. Lett. **68**, 2475 (1996)
[7.139] R. Bachelot, A. Lahrech, P. Gleyzes, A.C. Boccara: Proc. SPIE Int. Soc. Opt. Eng. (USA), Vol. 2782, 535 (1996)
[7.140] W.E. Moerner, L. Kador: Phys. Rev. Lett. **62**, 2535 (1989)
[7.141] D. Oesterhelt, J. Tittor, E. Bamberg: J. Bioenergetics and Biomembranes **24**, 181 (1992)
[7.142] P. Schmidt, U.H. Westphal, K. Worm, S. Braslavsky, W. Gärtner, K. Schaffner: J. Photochem. Photobiol. B **34**, 73 (1996)
[7.143] W.E. Moerner, T. Plakhotnik, T. Irngartinger, U.P. Wild, D.W. Pohl, B. Hecht: Phys. Rev. Lett. **73**, 2764 (1994)
[7.144] J. Hwang, L.K. Tamm, C. Bohm, T.S. Ramalingam, E. Betzig, M. Edidin: Science **270**, 610 (1995)
[7.145] M.H.P. Moers, N.F. van Hulst, G.T. Ruiter, B. Bölger: Ultramicrosc. **57**, 298 (1995)
[7.146] A. Jalocha, N.F. van Hulst: J. Opt. Soc. Am. B **12**, 1577 (1995)
[7.147] D.A. Higgins, J. Kerimo, D.A. Vandenbout, P.F. Barbara: J. Am. Chem. Soc. **118**, 4049 (1996)
[7.148] E. Betzig, R.J. Chichester, F. Lanni, D.L. Taylor: Bioimaging **1**, 129 (1993)
[7.149] R. Maheshwari, H. Tatsumi, Y. Katayama, M. Ohtsu: Opt. Commun. **120**, 325 (1995)
[7.150] M.H.P Moers, A.G.T. Ruiter, N. F. van Hulst: Ultramicrosc. **61**, 279 (1995)
[7.151] A. Kirsch, C. Meyer, T.M. Jovin: Integration of optical techniques in scanning probe microscopes: in *Analytical Use of Fluorescent Probes in Oncology*, ed. by E. Kohen, J.G. Hirschberg, NATO ASI Series, Life Sciences, Vol. 286 (Plenum, New York 1996)
[7.152] P.J. Moyer, S.B. Kaemmer: Appl. Phys. Lett. **68**, 3380 (1996)
[7.153] D.A. Higgins, P.F. Barbara: J. Phys. Chem. **99**, 3 (1995)
[7.154] T. Kobayashi (ed.): *J-Aggregates* (World Scientific, Singapore 1996)
[7.155] A. Kramer, T. Hartmann, R. Eschrich, R. Guckenberger: Ultramicrosc. (submitted)

[7.156] H.F. Hess, E. Betzig, T.D. Harris, L.N. Pfeiffer, K.W. West: Science **264**, 1740 (1994)
[7.157] K. Brunner, G. Abstreiter, G. Böhm, G.Tränkle, G. Weimann: Appl. Phys. Lett. **64**, 3320 (1994)
[7.158] D. Gammon, E.S. Snow, B.V. Shanabrook, D.S. Katzer, D. Park: Phys. Rev. Lett. **76**, 3005 (1996)
[7.159] J.K. Rogers, F. Seiferth, M. Vaez-Iravani: Appl. Phys. Lett. **66**, 3260 (1995)
[7.160] J. Liu, T.F. Kuech: Appl. Phys. Lett. **69**, 662 (1996)
[7.161] J.K. Leong, J. Mc Murray, C.C. Williams, G.B. Sinngfellow: J. Vac. Sci. Technol. B **14**, 3443 (1996)
[7.162] R.D. Grober, T.D. Harris, J.K. Trautman, E. Betzig, W. Wegscheider, L. Pfeiffer, K. West: Appl. Phys. Lett. **64**, 1421 (1994)
[7.163] T.D. Harris, D. Gershoni, R.D. Grober, L. Pfeiffer, K. West, N. Chand: Appl. Phys. Lett. **68**, 988 (1996)
[7.164] Y.Toda, K. Motonobu, M. Ohtsu, Y. Nagamune, Y. Arakawa: Appl. Phys. Lett. **69**, 827 (1996)
[7.165] T. Saiki, S. Mononobe, M. Ohtsu, N. Saito, J. Kusano: Appl. Phys. Lett. **67**, 2191 (1995)
[7.166] S.K. Buratto, J.W.P. Hsu, J.K. Trautman, E. Betzig, R.B. Bylsma, C.C. Bahr, M.J. Cardillo: J. Appl. Phys. **76**, 7720 (1994)
[7.167] U. Ben-Ami, N. Tessler, N. Ben-Ami, R. Nagar, G. Fish, K. Lieberman, G. Eisenstein, A. Lewis, J. M. Nielsen, A. Moeller-Larsen: Appl. Phys. Lett. **68**, 2337 (1996)
[7.168] S.K. Buratto, J.W.P. Hsu, E. Betzig, J.K. Trautman, R.B. Bylsma, C.C. Bahr, M.J. Cardillo: Appl. Phys. Lett. **65**, 2654 (1994)
[7.169] M.S. Ünlü, B.B. Goldberg, W.D. Herzog, D. Sun, E. Towe: Appl. Phys. Lett. **67**, 1862 (1995)
[7.170] J.W.P. Hsu, E.A. Fitzgerald, Y.H. Xie, P.J. Silverman: Appl. Phys. Lett. **65**, 344 (1994)
[7.171] T. Saiki, N.Saito, J.Kusano, M. Ohtsu: Appl. Phys. Lett. **69**, 644 (1996)
[7.172] A. Richter, M. Süpitz, Ch. Lienau, T. Elsässer, M. Ramsteiner, R. Noetzel, K.H. Ploog: Surf. Interface Analysis **25**, 583 (1997)
[7.173] R. Noetzel, M. Ramsteiner, J. Menniger, A. Trampert, H.P. Schoenherr, L. Deweritz, K.H. Ploog: Jpn. J. Appl. Phys. **35**, L297 (1996)
[7.174] G. Krausch, S. Wegscheider, A. Kirsch, H. Bielefeldt, J.C. Meiners, J. Mlynek: Opt. Commun. **119**, 283 (1995)
[7.175] A. Naber, H. Kock, H. Fuchs: Scanning **18**, 576 (1996)
[7.176] H.P. Zingsheim: Ber. Bunsenges. Phys. Chem. **80**, 1185 (1976)
[7.177] J. Calvert: J. Vac. Sci. Technol. B **11** 2157 (1993)
[7.178] S. Jiang, J. Ichibashi, H. Monobe, M. Fiujihira, M. Ohtsu: Opt. Commun. **106**, 173 (1994)
[7.179] H.J. Mamin, D. Rugar: Appl. Phys. Lett. **61**, 1003 (1992)
[7.180] H.J. Mamin, D. Rugar: Appl. Phys. Lett. **69**, 433 (1996)
[7.181] J. Jersch, K. Dickmann: Appl. Phys. Lett. **68**, 868 (1996)
[7.182] J. Jersch, F. Demming, K. Dickmann: Appl. Phys. A **64**, 29 (1997)
[7.183] B.D. Terris, H.J. Mamin, M.E. Best, J.A. Logan, D. Rugar, S.A. Rishton: Appl. Phys. Lett. **69**, 4262 (1996)
[7.184] U.Ch. Fischer, H.P. Zingsheim: J. Vac. Sci. Technol. **19**, 881 (1981)
[7.185] H.W. Deckman, J.H. Dunsmuir: Appl. Phys. Lett. **41**, 377 (1982)

[7.186] A.C. Pease, D. Solas, E.J. Sullivan, M.T. Cronin, C.P. Holmes, S.P.A. Fodor: Proc. Nat'l Acad. Sci. USA. **91**, 5022 (1994)
[7.187] S. Palacin, P.C. Hidber, J.-P. Bourgoin, C. Miramond, C. Fermon, G.M. Whitesides: Chem. Mater. **8**, 1316 (1996) and references therein
[7.188] E. Betzig, M. Isaacson, A. Lewis: Appl. Phys. Lett. **51**, 2088 (1987)

Subject Index